M. Samii · E. Knosp

Approaches to the Clivus

Approaches to No Man's Land

With 123 Figures in 193 Separate Illustrations

Springer-Verlag

Berlin Heidelberg New York
London Paris Tokyo
Hong Kong Barcelona
Budapest

Professor Dr. med. Prof. h.c. MADJID SAMII
Medizinische Hochschule Hannover
Direktor der Neurochirurgischen Klinik
am Krankenhaus Nordstadt
der Landeshauptstadt Hannover, Haltenhoffstraße 41,
W-3000 Hannover 1, FRG

Professor Dr. med. ENGELBERT KNOSP
Neurochirurgische Universitätsklinik
Währinger Gürtel, 18–20, A-1090 Wien
Present address: Neurochirurgische Klinik und Poliklinik
Johannes-Gutenberg-Universität Mainz
Langenbeckstraße 1
W-6500 Mainz, FRG

ISBN-13:978-3-642-76616-9   e-ISBN-13:978-3-642-76614-5
DOI: 10.1007/978-3-642-76614-5

Library of Congress Cataloging-in-Publication Data. Samii, M. (Madjid) Approaches to
the clivus–approaches to no man's land / M. Samii, E. Knosp ; with figures in separate
illustrations. p. cm. Includes bibliographical references and index. ISBN-13:978-3-642-7-
6616-9   1. Brain–Cancer–Surgery. 2. Skull base–Surgery. I. Knosp, E.
(Engelbert) II. Title. RD663.S36 1990 91-34436 616.99′481059–dc20 CIP

Reproduction of the figures: Gustav Dreher GmbH, W-7000 Stuttgart, FRG
Typesetting: Konrad Triltsch, Graphischer Betrieb, W-8700 Würzburg, FRG
11/3130-5 4 3 2 1 0 – Printed on acid-free paper

Preface

Tumors of the clivus have always been an unsolved problem for neurosurgery; they have in part remained so until today. Results are often discouraging in spite of the fact that such tumors are histologically mostly benign. In spite of, or because of all this, these tumors represent a great challenge to neurosurgery today. Given the speed at which diagnostic technologies develop, we have a headstart over the pioneers of neurosurgery – who for the most part considered tumors of this localization inoperable – and we must make use of our lead.

The importance of computed axial tomography (CT), including the possibilities of coronal-plane images and reconstruction as well as three-dimensional reconstruction of the skull, cannot be estimated highly enough. Not only can we gather a large number of details, but lesions are also exposed from different angles corresponding to the various approaches. These options enable the surgeon to weigh the advantages and disadvantages of each approach against the others with great accuracy and facilitate reasoning in the preoperative stage. Especially in neuroradiology, magnetic resonance imaging (MRI) has come to the fore alongside CT and will definitely complement other diagnostic techniques with many details. CT- or MRI-guided stereotactic biopsies, which at present are being increasingly employed in preoperative diagnosis of cerebral tumors, may also be used in diagnosing tumors of the skull base. When the histology of the tumor is known, the approach to use in the operation can be planned with greater accuracy.

Digital subtraction angiography and supraselective angiography offer the surgeon a wide range of new information about the vascularization pattern, which is of great neurosurgical interest. Apart from that, endovascular techniques in general, and preoperative embolization in particular, have developed into a new tool for the neurosurgeon.

These new technological modalities have not only made possible more accurate description of the lesion, but have also greatly improved early diagnosis in most cases. Such modalities, however, force us to develop correspondingly refined operative techniques, involving not only microsurgical techniques, but also the search for improved approaches which are in each case appropriate to the lesion in question.

All the approaches described in this book, and their variations, have one common target: the clivus. In many cases, it will only be a potential target, as the lesion will be located "on the way" and will be treated sufficiently before the clivus is reached. Nevertheless, we believe that it is justifiable to regard the topographical center of the skull base as the common target. Of course, we do not want to imply that "all roads lead to Rome", we should rather like to take the opportunity for open discussion on the most significant advantages and drawbacks of the approaches described. In our opinion, apart from the topographical extensions of the various tumors, the issue of their pathology is of decisive importance. When we know what type of tumor is involved or in all probability to be expected, we are usually able to delineate the surrounding structures from experience.

The relation of the lesion in question to the dura is of great significance for the choice of the approach, for various reasons. Firstly the dura forms a mechanical barrier protecting delicate intradural structures in cases of extradural lesions, and this is particularly true of the compact clival dura. Secondly, basal brain tumors are supplied by dural vessels, a fact which is of considerable significance for surgical strategies, especially if extensive vascularization is expected.

Since these approaches often cross the traditional boundaries of specialties, we favor *intraoperative interdisciplinary cooperation;* after all, learning is usually most effective and advantageous in such situations. We believe that the dura is not a boundary separating special fields, but an invitation to collaboration. Many roads lead to the clivus – to "no man's land."

We would like to express our gratitude, first of all, to Prof. K. Schürmann, whose thoughts, efforts, and achievements paved the way for the Skull Base Study Group's commitment to deal with the special problems of the cranial base. This interdisciplinary group has not only addressed special aspects or problems of surgery but has served as example on how the dura, as a symbolic barrier between specialties, can be overcome. We especially thank the anatomists Prof. J. Lang and Prof. P. Rabischong, who have contributed to bringing together the other specialties on a common ground, guiding us through the complexity of this region. Neuroradiologists Prof. J. Vignaud and Prof. P. Lasjaunias have served as active link, literally "showing" us the required details and offering new therapeutic options for lesions involving the head and neck. Without their work, many of the achievements made in recent years would not have been possible. Special recognition must also go to Prof. W. Draf and our ENT colleagues who have so often, hand in hand, collaborated intraoperatively.

The many years of friendship and professional bonds between Vienna and Hannover have made this work possible. Prof. W. Th. Koos and Prof. W. Firbas merit our special recognition.

And last but not least we thank Mrs. I. Dobsak for her excellent illustrations, Mrs. R. Wallisch for the exact iconography, and Mrs. E. Weber for her detailed work on the manuscript.

To all who have assisted us we express once again our gratitude.

M. SAMII · E. KNOSP

Contents

Introduction: The Clivus

The central portion of the skull base from the posterior part of the sella turcica to the foramen magnum was first named "clivus" by Johannes Blumenbach. This is a purely morphological description; the formation is ontogenetically heterogenous. The part of the skull base termed "clivus" develops as the body of the sphenoid bone fuses with parts of the occipital bone (basioccipital bone). Until the onset of puberty, these bones are separated by a synchondrosis, so the relatively homogeneous aspect only develops at an adult stage. However, the first problems relating to this morphological description arise as soon as one attempts to delineate the surrounding structures.

Osseous Boundaries

Seen from the interior aspect of the skull, it is easiest to define the cranial boundary of the clivus by the dorsum sellae and its caudal boundary by the anterior margin of the foramen magnum. The marked occipitopetrosal suture and the jugular foramen mark the boundary between the petrous bone and the clivus. It is difficult to define the boundary between the clivus and the occipital squama, which runs along the junction of the jugular foramen and the posterior rim of the occipital condyle. Thus, the anterolateral and posterolateral parts of the occipital bone, which fuse at an early stage of development, are incorporated in the clivus. The sphenoidal part of the clivus is delineated by the dorsum sellae and the posterior wall of the sphenoidal sinus. Demarcation against the remaining body of the sphenoid bone is impossible, as the delineation of the clivus in this region is only hypothetical. The ventrocaudal portion of the clivus borders on the epipharynx at the exterior skull base. The above "outline" has in itself already provided a survey of the regions from which lesions may encroach on the clivus and from which the surgical approach must start.

The clivus is basically a cancellous bone with inner and outer tables. It is therefore suited to osteosynthesis (see Chap. 1).

Table 1. Dimensions of the osseous clivus[a]

Length:	42–54 mm
Width:	28 mm (measured in the center of the clivus)
Sphenoidal-clival angle:	116° (96–143°)[b]

[a] Source: v. Lanz and Wachsmuth 1979
[b] Lang 1981

Dural Boundaries

The dura of the clivus is part of the basal envelope of the posterior fossa, which is in continuity with the dura of the posterior aspect of the petrous bone as well as the spinal canal.

In the cranial portion around the dorsum sellae, delineation becomes very difficult due to the complex structure of the tentorial notch. The posterior petroclinoidal fold stretches from the dorsum sellae to the tentorium. The anterior petroclinoidal fold, which starts at the anterior clinoid process, forms the tentorial notch. Thus, the so-called *Wannenregion* – the "trough" between the two dural folds – directly adjoining the dorsum sellae forms the roof of the cavernous sinus.

In this area, the third and fourth cranial nerves enter the dura of the lateral wall of the cavernous sinus. The clival dura itself is extraordinarily compact and difficult to detach from the bone. This is particularly marked in children and young adults in the region of the spheno-occipital synchondrosis.

The two dural layers are interconnected by numerous septa. In the region of the dorsum sellae the dura bridges the posterior drainage of the cavernous sinus into the inferior petrosal sinus. The latter joins the cavernous sinus to the jugular foramen and represents the lateral boundary between the clivus and the petrous bone. However, in only 45% of all cases considered does the appearance of the inferior petrosal sinus correspond to the textbook (Shiu et al. 1968). The more numerous the connections to the network of the basilar plexus, the smaller the inferior petrosal sinus. In 25% of cases the inferior petrosal sinus is entirely missing, so that the cavernous sinus is connected to the veins of the anterior condylar canal and the prevertebral venous plexus. According to Renn and Rhoton (1975), the network of the basilar plexus has developed in 82% of all cases considered; it consists of an irregular venous plexus between the two dural layers with numerous connections to the inferior petrosal sinuses. At the transitional zone between the cavernous and the inferior petrosal sinus in the level of the dorsum sellae we frequently find a pronounced connection to the contralateral side, known as the dorsum sellae sinus. The lateral part of the dorsum sellae and the apex of the petrous bone are attached to a ligament, the superior sphenopetrous ligament (Gruber's ligament). The ligament and the posterior part of the sella turcica form a triangular lacuna, Dorello's canal, which transmits the sixth cranial nerve in its lateral part (Lang 1975; Nathan et al. 1974; Umanski et al. 1991).

Ligaments of the craniocervical junction are attached to the most caudal portion of the clivus: the membrana tectoria, the cranial parts of the cruciate ligament, and the apical ligament of the odontoid process (see Fig. 1.1).

Arterial Supply of the Clivus

There are two main feeding vessels of the basal dura in the clival region, the meningohypophyseal trunk, branching off the internal carotid artery, and the posterior meningeal artery, originating from the ascending pharyngeal artery. The clivus is located at the border between the supply regions of two arteries, so many variations and combinations are possible. This is of particular signif-icance in cases where tumors have encroached on the clivus, because the vascular supply adopted is that at the site of origin; this is an important factor for the planning of surgical interventions.

The meningohypophyseal trunk of the internal carotid artery and its clival branches enter the clivus from cranially. A medial group of branches forming an anastomosis with the contralateral side may also be exposed; this supplies the upper clivus and stretches as far as the point where the abducent nerve enters the dura. A lateral clival artery passes forward along the inferior pe-trosal sinus and in its course forms an anastomosis with the posterior meningeal artery.

The lateral clival artery may, in rare instances, also orginate in the intra-cavernous segment of the internal carotid artery as a branch in its own right (Parkinson 1964; Lang and Schäfer 1976; Lasjaunias et al. 1977; Knosp et al. 1987). The neuromeningeal trunk branch off the ascending pharyngeal artery supplies the cranial nerves conveyed by the jugular foramen and, as the poste-rior meningeal artery, enters the posterior cranial fossa via the jugular fora-men. A second branch supplies the hypoglossal nerve and enters the posterior fossa through the hypoglossal canal. It supplies the dura in the lowest part of the clivus, the anterior margin of the foramen magnum, and the craniocervical junction (Fig. 1). Covered by the sigmoid sinus, a meningeal branch of the occipital artery enters the posterior fossa via the mastoid emissary; it forms an anastomosis with the posterior meningeal artery. The dorsal aspects of the foramen magnum are supplied by a dural branch of the vertebral artery, which originates at the dural entry area of the vertebral artery.

Development

Osseous Clivus

In terms of development, the clivus is not a homogeneous formation. It con-sists of the body of the sphenoid bone and the basilar part of the occipital bone.

These two osseous parts remain separated by the spheno-occipital synchondrosis into adolescence (up to the age of 12 years according to Torklus and Gehle 1987). The synchondrosis is situated at the transitional zone between the upper and middle third of the clivus. Ossification at the synchondrosis sets in from the inside of the skull. The cartilagenous formation may be the site of origin of clival chondroma or chondrosarcoma. Neither is the occipital bone itself a homogenous bone; it consists of several portions. The basilar part of the occipital bone, which is connected to the sphenoid bone, is on both sides adjoined by an anterior lateral part (exoccipital) and a posterior lateral part. The two latter formations are separated by the interoccipital synchondrosis, which passes through the occipital condyle and ossifies at an earlier age than the spheno-occipital synchondrosis (at the age of 6 years according to Lang 1981). The interoccipital synchondrosis marks the course of the hypoglossal canal which runs between the two osseous parts of the occipital bone and is covered with bone during ossification. The hypoglossal canal may be double (Lang 1986a), which is regarded as a rudimentary form of the vertebral formations fusing to form the occipital condyle (Torklus and Gehle 1987). The occipital condyle owes its shoesole-like look to the fact that it develops from two parts.

Due to different forms of ossification, the squamous part of the occipital bone consists of two parts: the nuchal part, which borders on the foramen magnum, and the occipital part. A persistent fissure or suture between these two parts at the level of the transverse sinus is referred to as the sutura mendosa (Lang 1981). Transverse fissures rarely develop in the clival region. Such rudiments are referred to as "basilar transverse fissures"; they are believed to be manifestations of the vertebrogenic development of the basioccipital bone (Wackenheim 1985).

Craniocervical Junction

After complete fusion of the primary vertebrae located furthest cranially, the atlas represents the fifth primary vertebra (Albrecht as quoted by Torklus and Gehle 1987). The odontoid process of the axis represents the vertebral body of the atlas, which fuses with vertebral body C2. Until the age of 3 years these two formations remain separated by the subdental synchondrosis, the vestiges of the notochord between C1 und C2. In terms of development, the anterior arch of the atlas is the sole identifiable manifestation of a hypochordal arch. The proatlas, a vertebra which in terms of development is located further cranially than the atlas, is normally no longer manifest. Its vertebral body is incorporated in the apex of the odontoid process as an osseous core in its own right. It may, however, also be found in an isolated form, in which case it is referred to as Bergmann's terminal ossicle. In disadvantageous cases, e.g., in cases of concurrent odontoid hypoplasia and loosened ligaments, such a malformation may lead to complaints (see Chap. 1). Normally, the terminal ossicle is incorporated into the apex of the odontoid process by the age of 12 years. If it is not, it is referred to as the "odontoid bone." One must, however, distinguish between an actual odontoid bone (vertebral body C1 with hypoplasia of the odontoid process) and the "so-called odontoid bone" which in terms of formation is part of the proatlas.

A vertebra situated short of the proatlas, called the ante-proatlas, is believed to be responsible for the development of a third condyle (Putz 1975). This protuberance, situated at the median aspect of the anterior margin of the foramen magnum, articulates with the odontoid process of the axis, and occasionally also with the arch of the atlas. On the other hand, vertebral formations that normally remain separate may also fuse. The most commonly found example of this is assimilation of the atlas, which may also develop unilaterally. Atlas assimilations are found in less than 1% of all cases considered (v. Lanz and Wachsmuth 1979). Such assimilations of the atlas may occur alone or, in various forms, concurrently with other malformations of the craniocervical junction, but they are pathological in the latter case only.

Vascularization

At an early embryonic stage, anastomoses link the internal carotid and vertebral arteries (Padget 1948). The primitive trigeminal artery connects the posterior loop of the carotid artery (the C5 segment of the internal carotid artery) and the basilar artery. The lateral clival branch of the meningohypophyseal trunk is believed to have developed from this anastomosis (Lasjaunias 1981). Other authors, however, claim that the primitive trigeminal artery exits the internal carotid artery separately from the meningohypophyseal trunk (Lang in a conversation with the authors; Parkinson and Schields 1974; Tschabitscher 1990). Diffentiations between various types are made according to the area of supply in the vertebrobasilar system (Lasjaunias 1981; Krayenbühl and Yaşargil 1979). The primitive trigeminal artery is not pathological in itself, it may, however, be the site of origin of aneurysms (Krayenbühl and Yaşargil 1979; Kerber and Manke 1983; Morita et al. 1989).

The primitive otic artery is a carotid-basilar anastomosis which only persists in extraordinarily rare cases (Lasjaunias 1981). More often, a primitive hypoglossal artery persists; this passes forward through the hypoglossal canal and connects the internal carotid artery to the contralateral vertebral artery. There are two further connections between segments of the carotid and vertebral arteries in the form of proatlantic arteries. Type 1 runs from the internal carotid artery to the vertebral artery between C0 and C1, while type 2 communicates between the external carotid artery and the vertebral artery between C1 and C2.

The development of the venous pattern of the skull base is even more complex than the comparatively constant arterial system. The inferior petrosal sinus develops from the ventral myeloencephalic vein and communicates with the cavernous sinus, which develops from the pro-otic sinus. This venous system connects the supraorbital and maxillar veins with the internal jugular vein through the cavernous and inferior petrosal sinuses (Padget 1956). The inferior petrosal sinus and the jugular vein originally join outside the skull base. The basilar plexus, which is a continuation of the prevertebral venous plexus, has several anastomoses with the inferior petrosal sinus, the veins of the hypoglossal canal, the jugular vein and the venous system of the foramen magnum. There is still much discussion about the question of whether the cavernous sinus is a sinus traversed by trabeculae (Harris and Rhoton 1976) or a plexus of veins (Parkinson 1964; Taptas 1982). Examination of fetuses (Knosp et al. 1987) supports the theory that the cavernous sinus is a plexus of veins surrounding the internal carotid artery and partly draining into the inferior petrosal sinus. Apparently, the fact that the venous structures are difficult to trace in adults is due to postnatal changes in the skull base. Another point to stress is that the adult venous pattern has not yet completely developed (or completely dwindled) at birth (Knosp et al. 1987).

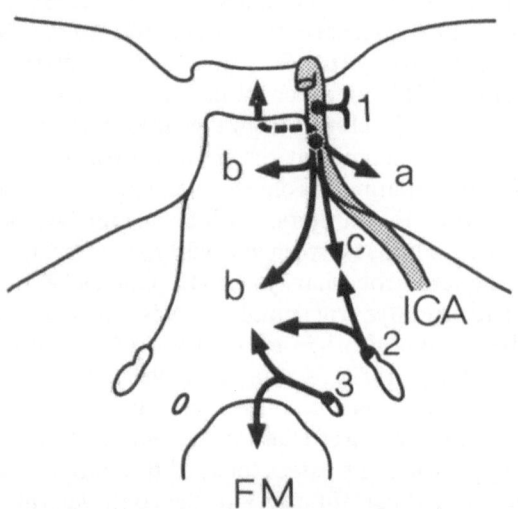

Fig. 1. Vascularization of the clival dura. The blood supply of the clival dura is by the meningo-hypophyseal trunk, which originates from the internal carotid artery (*ICA*) and bifurcates directly afterwards, as well as by branches of the ascending pharyngeal artery, namely the posterior meningeal arteries, which reach the dura via the jugular foramen and the hypoglossal canal. Normally, the inferior lateral trunk (*1*) does not play a part in the supply of the clivus. The meningohypophyseal trunk and its branches are shown: inferior hypophyseal artery (*dotted line*), tentorial artery (*a*), medial clival branches (*b*), and lateral clival branch (*c*). Branches of the ascending pharyngeal artery ramify through the jugular foramen (*2*) and the hypoglossal canal (*3*). *FM*, foramen magnum

Chapter 1: Transoral Approach

Historical Survey

The development of this approach to the clivus was triggered by favorable results in transoral incisions of retropharyngeal tuberculous abscesses. In 1957, Southwick and Robinson, who were surprised at the low rate of complications caused by infections, reported on the successful removal of an osteoma of vertebral body C2; later, Fang and Ong (1962) and Fang et al. (1964) gave notice of a series of transoral operations for atlantoaxial dislocations of various origins.

Mullan et al. made decisive progress in 1966 when operating on a lesion of the clivus which had also encroached on the dura (a fibrosarcoma, to be exact). They described very well the view of the basilar artery and the vertebral junction they were affered this way. Naturally, mid-basilar artery aneurysms were next among the great challenges to neurosurgery. Sano et al. (1966), Drake (1969), and Yaşargil (1969, 1970) were the first to venture upon such operations. They reported successful transoral clipping of a basilar artery aneurysm. Subsequently, Drake and Yaşargil abandoned this method because of the overly high risk of infections (Drake 1973; Peerless and Drake 1982; Yaşargil 1984).

In spite of the high risk of infections and the unconventionality of the approach, there were a number of further reports on transoral clippings of basilar artery aneurysms (Hashi et al. 1976; Laine-Jomin 1977, Jomin and Bouasakao 1977; Saito 1978; Matricali and Dulke 1981; Hayakava et al. 1981). Most authors described successful transoral operations on extradural lesions, such as craniocervical malformations, C1–C2 subluxations and basilar invaginations (Greenberg 1968, Greenberg et al. 1968; Eldridge 1967; Sukoff et al. 1972; Van Gilder and Menezes 1977; Menezes et al. 1980, 1985; Fox and Jerez 1977; Delandsheer et al. 1977; Derome 1977; Derome et al. 1977; Apuzzo et al. 1978; Pasztor et al. 1980; Di Lorenzo 1982) as well as chordomas of the clivus (Guthkelch and Williams 1972; Krayenbühl and Yaşargil 1975; Wood et al. 1980; Delgado 1981; Pasztor et al. 1984; Pasztor 1985). But also neurinomas have been removed by this route (Crockard and Bradford 1985; Crockard et al. 1991).

Anatomy

Anatomy Relevant to Surgery

The *soft palate* consists of an aponeurosis attached to the hard palate, into which muscles project (the levator and tensor veli palatini muscles) and which gives attachment to muscles proceeding to the tongue and pharynx (the palatoglossal and palatopharyngeal arches). The soft palate may be split along its median raphe without interrupting blood supply through the descending palatine artery.

The posterior wall of the *pharynx* consists of mucosa, submucosa, and the pharyngeal constrictors, which are easy to dissect from the prevertebral muscles. The prevertebral muscles below are attached to the lower surface of the clivus on both sides of the pharyngeal tubercle, which is situated at a distance of 11 mm from the anterior margin of the foramen magnum. The posterior wall of the pharynx is supplied by the ascending pharyngeal artery, which proceeds medial to the internal carotid artery along the lateral rim of the rectus capitis anterior muscle. The anterior arch of the atlas, together with the anterior tubercle are easily palpable through the posterior wall of the pharynx.

The *anterior longitudinal ligament* is attached to the arch of the atlas from caudally and is continued in the atlanto-occipital membrane, which passes forward from the arch of the atlas to the clivus (Fig. 1.1).

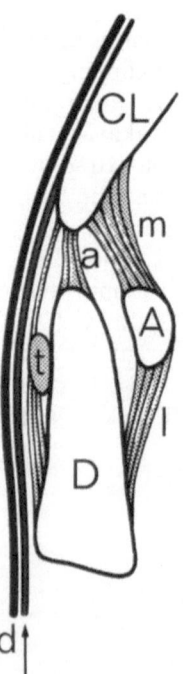

Fig. 1.1. Ligaments of the craniocervival junction (median section): The cone-shaped alar ligaments originate from the medial and anterior rim of the atlanto-occipital joint and run to the lateral and anterior surfaces of the dens axis (not visible in the median section). The transverse ligament of the atlas (*t*) proceeds from the tubercles at the internal aspect of the lateral mass of the atlas. It is the strongest ligament at the craniocervical junction (2 mm thick and 10 mm wide), and adjoins the odontoid process of the axis dorsally. Together with the alar ligament, it prevents the odontoid process from moving backward when the head is bent. The apical ligament of the odontoid process (*a*) passes forward from the foramen magnum to the apex of the odontoid process. It is situated between the alar ligaments and is regarded as the vestige of the notochord. Thinly developed longitudinal fascicles proceed from the posterior surface of C2 to the transverse ligament and on to the posterior rim of the clivus, thus creating the impression of a cruciate ligament at the posterior aspect of the odontoid process when seen from dorsally. The posterior longitudinal ligament thickens as it is continued in the tectorial membrane (*arrow*); it enters the clivus in the lower half. In the cranial part it is fused with the clival dura (cf. Figs. 1.4, 1.5). *CL*, Clivus; *D*, odontoid process; *A*, anterior rim of the atlas; *l*, anterior longitudinal ligament coming from caudally; *m* anterior antlanto-occipital membrane; *D* dura

Table 1.1. Dimensions and topographical relationships[a]

Distances between:

Carotid arteries	50 mm (42–60 mm)
At the skull base:	
Hypoglossal canals	28 mm
Atlanto-occipital articulations	23 mm (16–30 mm)
Angle between the hypoglossal canal and the median plane:	45°
Odontoid process	
Length:	15 mm
Sagittal diameter:	10.5 mm

Diameter of the myelon (mm):

	Sagittal	Transverse
Pontomedullary	15	18
C1	10.4	12.6
C2	9	10.2
C3	8	11

[a] Sources: Lang 1981; v. Lanz and Wachsmuth 1979

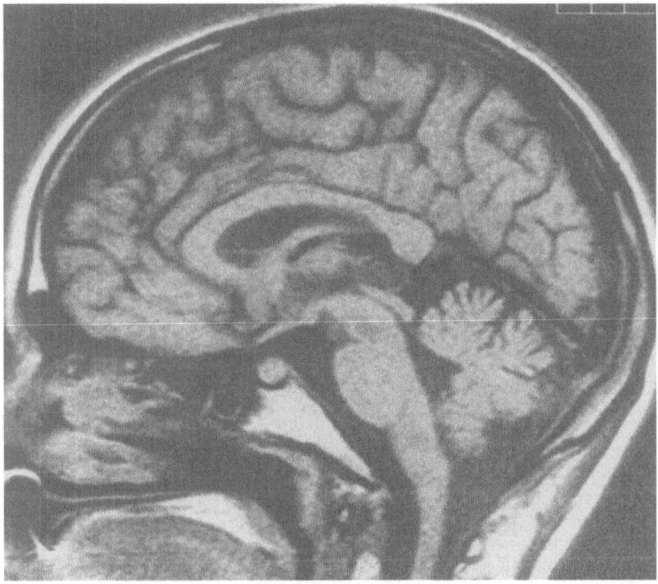

Fig. 1.2. Topographical relationships in transclival approaches. The junction of the lower and middle thirds of the clivus *corresponds to:* the pontomedullar groove and the junction of the vertebral arteries. The apex of the odontoid process *corresponds to:* the pyramidal decussation. The hard palate *corresponds to:* the level of the foramen magnum

Key Steps of the Approach

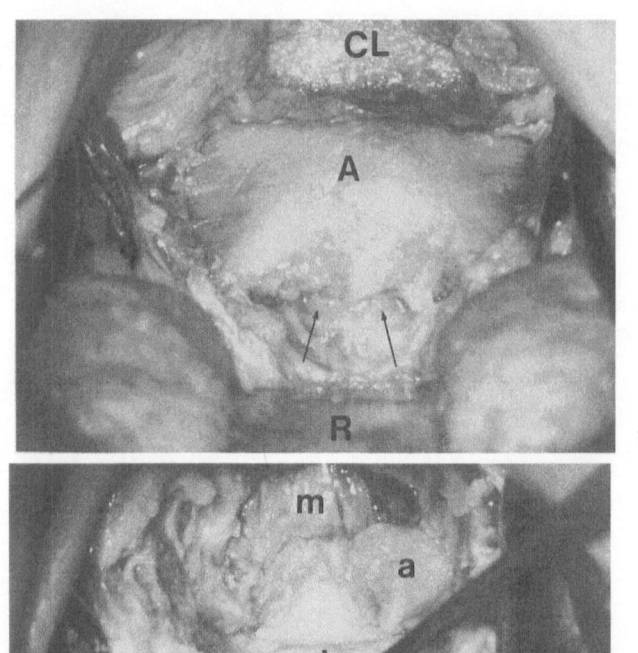

1.3

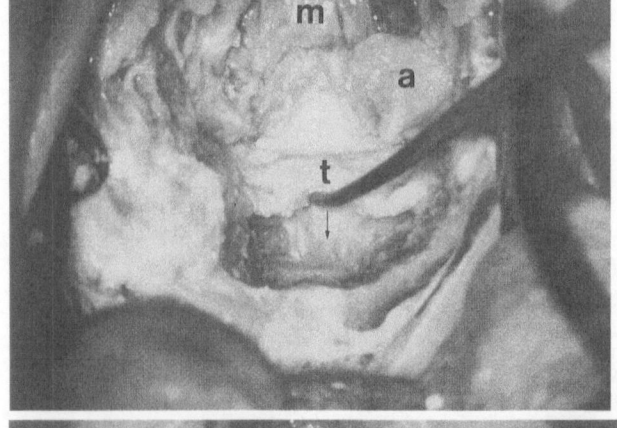

1.4

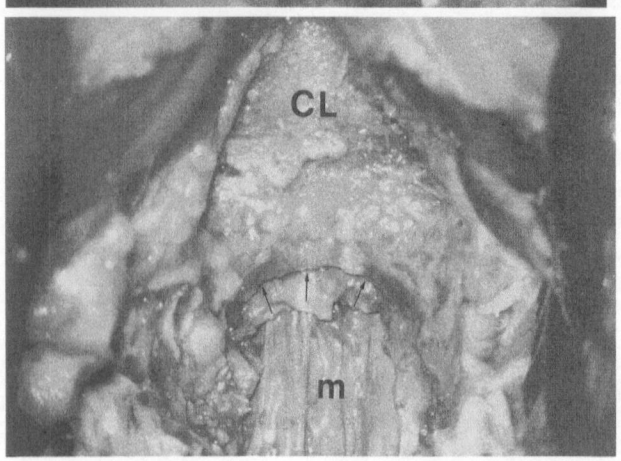

1.5

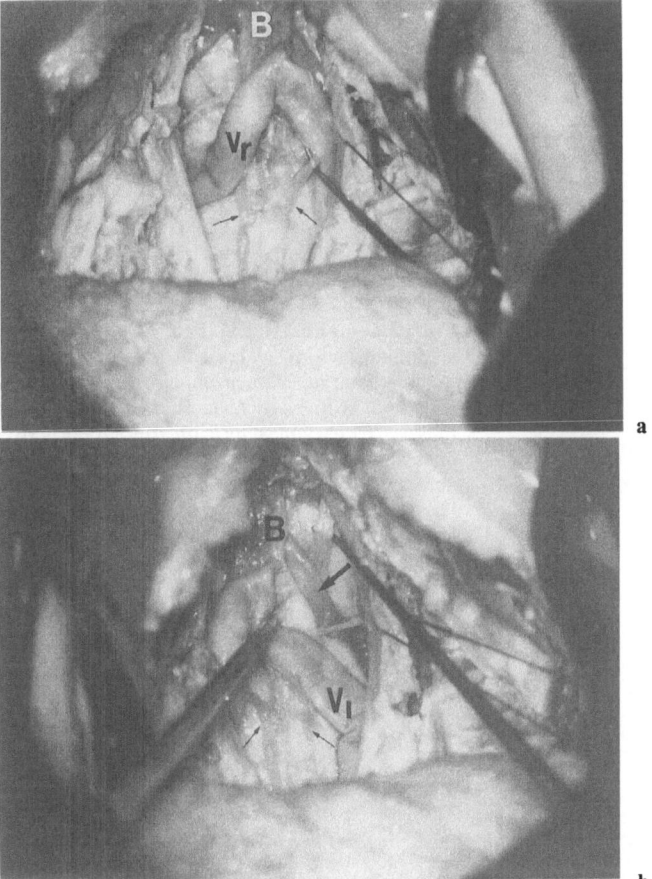

a

b

Fig. 1.6. a After opening the basal clival dura along the median, the vertebral junction may be exposed. The hook exposes the anterior spinal artery on the left. **b** The most cranial part of the clival craniotomy, affording a view of the basilar artery (*B*) beyond the exit of the inferior cerebellar artery (*arrow*). The abducent nerve on the left is held with a hook. V_r, right vertebral artery; V_l, left vertebral artery. *Small arrows* indicate anterior spinal arteries

Fig. 1.3. After the pharynx has been split along the median raphe and the rectus capitis anterior muscle has been dissected, the anterior arch of the atlas (*A*) is exposed. The tongue is depressed with a retractor (*R*). *CL*, inferior margin of the clivus. The base of the ondontoid process is visible below the arch of the atlas (*arrows*)

Fig. 1.4. The anterior rim of the arch of the atlas and the odontoid process have been resected. The caudal part of the cruciate ligament is exposed (*arrow*) as the transversa ligament is retracted with a dissector. See also Fig. 1.1. *a* Remains of the left alar ligaments; *m*, tectorial membrane; *t*, transverse ligament

Fig. 1.5. After the soft palate has also been split, the lower half of the clivus may be exposed (*CL*). *m* Tectorial membrane. The anterior margin of the foramen magnum is marked by *arrows*

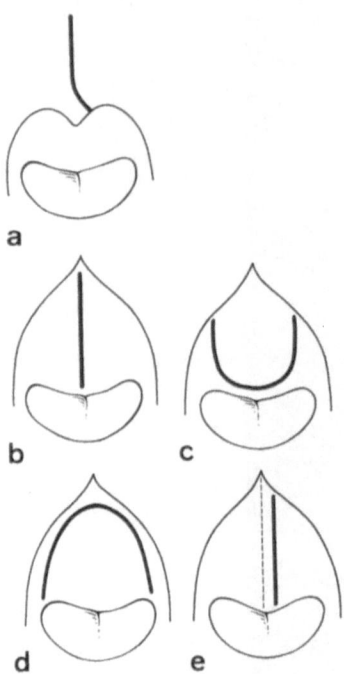

a

b   c

d   e

Fig. 1.7 a–e. Pharyngeal and palatal incisions. **a** Median incision of the soft palate sparing the uvula. **b** Pharyngeal incision according to Sano et al. (1966), with vertical incision of mucosa and muscles. **c** Pharyngeal incision after Drake (1969): horseshoe-shaped, petiolated upwards. **d** Incision according to Schmelzle and Harms (1986), with a large flap of mucosa, horseshoe-shaped and petiolated downward for possible bone grafts and osteosynthesis. **e** Paramedian mucosal incision and median muscle incision according to Pasztor (1984), so as to attain two-layered, safe wound closure

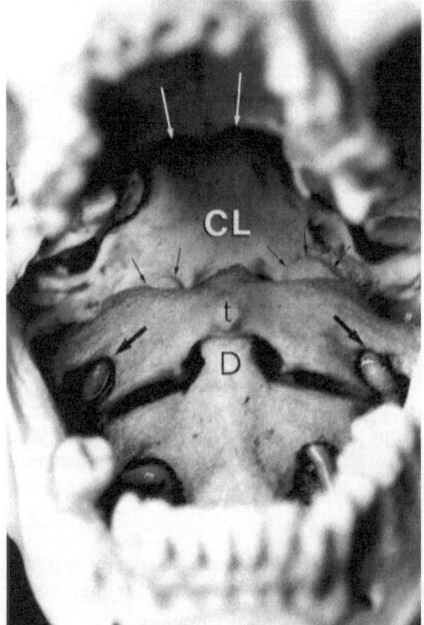

Fig. 1.8. View of the clivus and the upper cervical spine in a bone specimen. *CL,* Clivus; *t,* anterior tubercle of the arch of the atlas. *Small arrows* indicate the occipital condyle, *large arrows* the vertebral artery between C1 and C2, and *white arrows* the posterior rim of the hard palate

Surgery

Technique

Preparations start several days before surgery with local disinfection of the mouth and pharynx and a special antibiotic cover. The patient is orally intubated. Unless a Crutchfield extension is used in trauma cases, the head is reclined in a Mayfield headrest and lateral fluoroscopy is used for guidance. In critical cases, spinal cord monitoring is indicated from the start of intubation until the early postoperative stage to correct possible mistakes in positioning and observe the improvement of evoked potentials after decompression or to minimize operative traumatization of previously damaged medulla (Spetzler et al. 1979; Hardley and Spetzler 1986).

We consider splitting the soft palate inevitable in clival surgery; in surgery of the craniocervical junction not necessitated by malformations of the skull base, however, it is sufficient to retract the soft palate by means of a rubber band introduced through the nostril. Otherwise, the soft palate is infiltrated with local anesthetic and split along the median raphe. The two parts of the soft palate are then pulled aside like the wings of a double door by means of two rubber bands exiting through the nostrils, and raised to open the view of the upper pharynx. The pharynx is split in a vertical median incision (for various pharyngeal incisions see Fig. 1.7).

The prevertebral muscles are also split along the median and dissected subperiostally from the arch of the atlas, the anterior aspect of C2 and the clivus. Special care should be taken to preserve this layer in order to allow tight two-layered closure. Microsurgical techniques are used in drilling off the arch of the atlas and odontoid process to expose the anterior margin of the foramen magnum, the lower clivus, and the tumor. After endocapsular tumor resection, the tumor is excised in a piecemeal manner; demarcation of the adjacent bone is usually successful. Only then is the tumor removed from the dura.

Tumors invading the dura represent a basically different problem; surgery is definitely more extensive then. Firstly, the dura does not act as a separating and protective layer for the vertebral and basilar arteries, the perforating vessels of the latter, and the brain stem itself. Furthermore, it is very difficult to obtain watertight dural closure, due to space in the region being limited. Even in the absence of a dural defect, it is almost impossible to suture the clival dura because of its tightness. This is particularly true in pointed epipharynges, which are very common in case of malformation. If primary chronic polyarthritis necessitates transoral resection of the odontoid process, it is unsufficient to remove the osseous part alone. For an improvement of neurological functions, the fibrous ring must be transected and the granulation tissue removed.

Sufficient brain stem decompression has been attained only when the dura pulsates. We pay special attention to the closure of a possible dural defect. For watertight closure, we use fascia lata, muscle, and fibrin glue. In addition, external lumbar drainage is recommended until the wound has healed so as to prevent leakage of cerebrospinal fluid. The pharynx is closed in two layers in interrupted sutures. The same applies to the soft palate, in which proper approximation is very important to largely avoid phonatory impairment. The gastric tube is left in place for a few days for nutrition (NB: beware pressure ulcers of the pharyngeal wound).

Case Report

Outwardly, this 60-year-old patient gave the impression of a short-necked person. Active and passive mobility of the cervical spine were distinctly limited. During the past 6 years, progressive neurological signs had slowly evolved, starting with dysesthesia of both lower legs. Subsequently, tinnitus, vertigo, and paresis of the hypoglossal nerve on the left had set in. The patient suffered additionally from dysphagia and marked dysarthria. He was finally hospitalized because of progressive spastic tetraparesis. Plain skull radiographs revealed the morphological substrate of the neurological deficiencies to be a massive basilar impression. Depiction of the ventral compression at the brain stem was particularly impressive in MRI (Fig. 1.10).

After oral intubation, the patient was reclined in a Mayfield headrest. Positioning was monitored by means of somato-sensitive evoked potentials

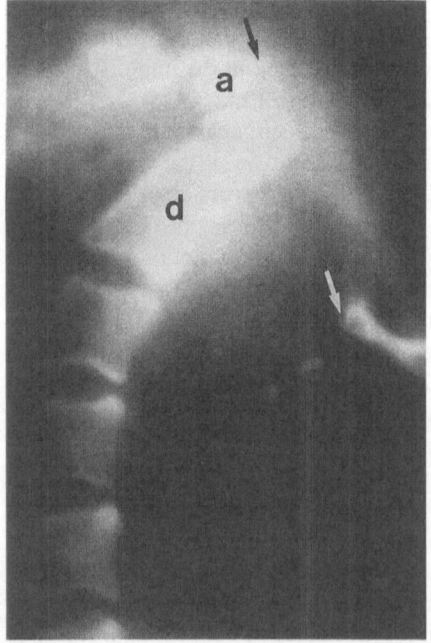

Fig. 1.9. Plain skull tomography showing severe basilar impression. *d*, Odontoid process; *a*, anterior arch of the atlas; *white arrow*, opisthion; *black arrow*, basion

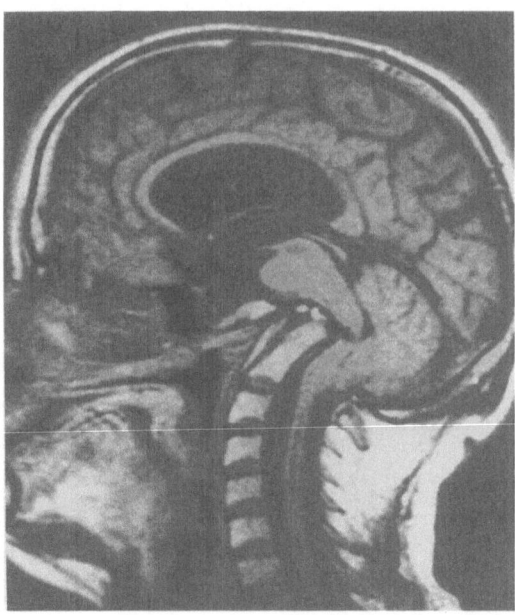

Fig. 1.10. MRI, median section, showing the enormous extent of the basilar impression. The epipharynx is drawn upwards like a pointed gable, the apex of the odontoid process is situated at the level of the dorsum sellae. The brain stem is deformed and extended over the ventral space-occupying lesion

(SEP). In addition, it was laterally guided by fluoroscopy (Fig. 1.13) and centered on the craniocervical junction. Fluoroscopy also assisted intraoperative observation of surgery.

Following normal disinfection of the nasopharynx, a combined retractor and tongue depressor was inserted; after infiltration, the soft palate was split along the median line. As exposure of the pointed pharynx was still insufficient after splitting of the soft palate, the incision had to be extended to the hard palate. The posterior part of the hard palate was removed and the posterior wall of the pharynx opened along the median line. The mucous membrane and underlying ventral cervical muscles were dissected subperiostally; this became increasingly difficult in the most cranial part of the epipharynx, which was affected by the malformation. The anterior arch of the atlas and the odontoid process were drilled off in a piecemeal manner with a diamond drill, until the ligaments of the craniocervical junction were reached and the basilar dura was exposed. Removal of the odontoid process and resection of the ligamentous part resulted in sufficient brain stem decompression. The pharynx was closed in interrupted sutures and the soft palate sutured in several layers. Postoperative swelling of the tongue, which we had been prepared for, was not the main problem, but the patient had to be sedated and ventilated for several days after surgery due to wound dehiscence in the soft palate. Left hypoglossal paresis, which had already been present before surgery, together with swelling and the abovementioned problems, made extubation difficult, thus necessitating temporary tracheotomy.

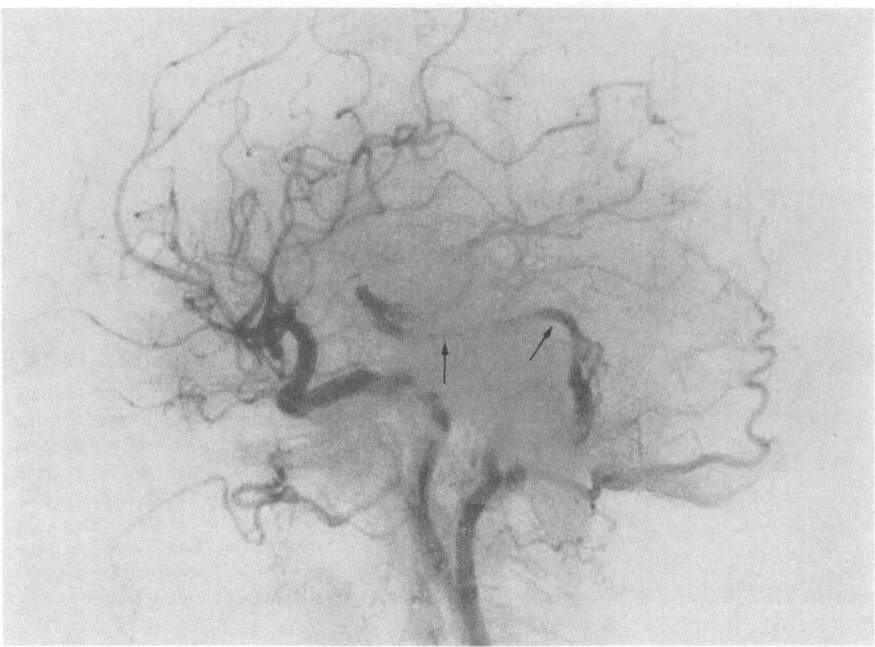

Fig. 1.11. Angiogram of the brachiocephalic trunk, with vertebral and basilar arteries displaced (*arrows*). The petrous segment of the internal carotid artery also has an abnormal course

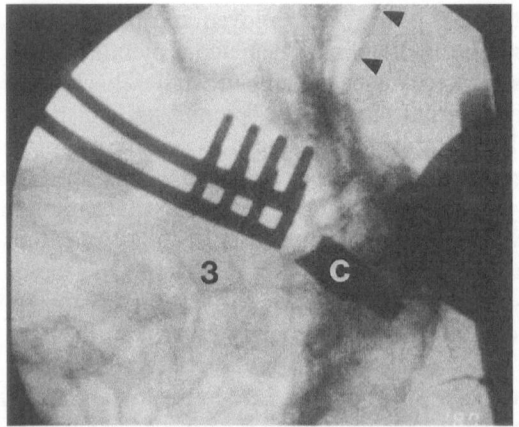

Fig. 1.12. Intraoperative fluoroscopy after resection of the odontoid process. The defect is filled with contrast medium (*c*) *3*, Vertebral body C3; *arrows*, anterior cranial fossa

When the patient was discharged into home care, tetraparesis had already started to improve. After 2 months, dysarthria was almost imperceptible and the gait pattern had further improved.

Frequently Encountered Lesions

First and foremost, *congenital, degenerative, and traumatic changes of the craniocervical junction* causing ventral compression of the medulla oblongata or the myelon are indications for transoral operation. The congenital abnormalities involve platybasia, isolated odontoid bones, or atlantal assimilations. In cases of traumatic origin it is mostly fractures of the odontoid process which have not fused that are causes for surgery. Degenerative diseases are the most frequent disorders of the craniocervical junction. Severe degenerative changes affect the joints and entail loosening and defective positions. They are classed together under the term "cranial settling" (Menezes et al. 1985). Degenerative changes of the atlanto-axial joint together with lowering of the lateral mass of the atlas result in elevation of the odontoid process and basilar impression. This entails "telescoping" of the anterior rim of the atlas over the odontoid process and at the same time ventral displacement of the posterior arch of the atlas (Fig. 1.13). This in turn often results in considerable medullary compression.

In extradural tumors of the lower clivus, in many cases *chordomas* or *chondromas*, the transoral approach constitutes an option for surgery. In such cases, how far the lesion extends laterally and toward the dorsum sellae is of paramount importance for surgical accessibility. The relation between tumor and dura affects the possibility of complications such as CSF leakage and possible subsequent meningitis.

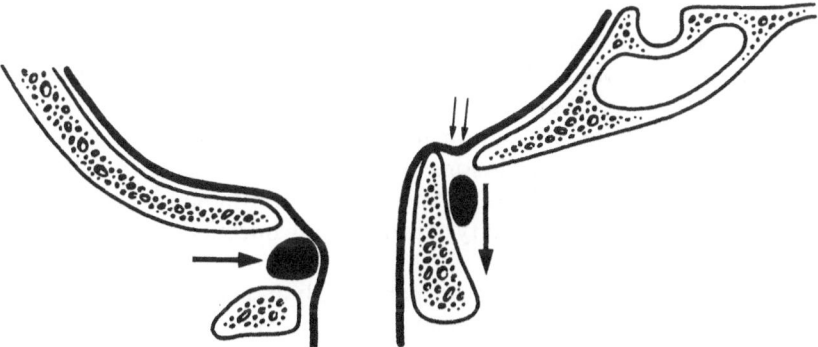

Fig. 1.13. Elevated position of the odontoid process and "telescoping" of the anterior arch of the atlas over the odontoid process (*vertical arrow*). This comes with dorsal compression of the myelon by the posterior rim of the atlas (*horizontal arrow*). The risk of damaging the dura is greatest between the clivus and the anterior rim of the odontoid process (*parallel arrows*)

Unlike in cases of malformation the anatomical conditions in a transoral approach to *aneurysms* of the middle segment of the basilar artery are normal. This facilitates the approach. Transdural surgery, however, involves the basic problem of watertight dural closure, which is almost impossible due to the compactness of the clival dura (Hardley and Spetzler 1986; Hardley et al. 1988).

A ventral approach to the basilar artery renders thorough examination of the vessel possible (Resch 1990), so fenestrations of the basilar artery as the site of origin of aneurysms can be observed (Matricali and Dulke 1981). Aneurysms always develop at the proximal spur of the fenestration (Hoffman and Wilson 1979), as was also shown in the illustrative work of Black and Ansbacher (1984).

According to von Mitterwallner (1955), fenestration or duplication of the basilar artery is observed in 0.3% of the cases reviewed, and fenestrations without duplication of the artery in 2% (Müller 1975, cited in v. Lanz and Wachsmuth 1979). Busch (1966) found 13 fenestrations of the basilar artery in 1000 brains examined. Most of the fenestrations were situated in the proximal segment of the basilar artery.

Advantages and Disadvantages of the Approach

Particularly because of the high risk of infection, we agree with Derome und Guiot (1979) that the transoral approach should primarily be used for treatment of extradural lesions. As is the case in other anterior approaches too, lateral space is very limited and the approach remains confined to the lower half of the clivus. Many authors regard the performance of tracheostomy beforehand as unavoidable (Greenberg 1968; Apuzzo et al. 1978; Pasztor et al. 1980; Menezes et al. 1980), but there are no surgical reasons that make this absolutely mandatory (Derome and Guiot 1979; Hardley and Spetzler 1986). Nevertheless, it may be necessitated by postoperative complications, as in the case we report. One argument against tracheostomy is that it constitutes a dangerous source of infection close to the pharyngeal wound.

Splitting the mandible, lip, and tongue as proposed by Wood et al. (1980) is not required for treatment of lesions of the clivus. Glossotomy would only be necessary for processes below C3. Special attention should be paid to the gastric tube because of the possibility of propagating infections and the danger of pressure necrosis of the wound.

Mullan et al. (1966) point out that, particularly in children, the adenoids may make it difficult to close the cranial part of the wound of the pharyngeal incision. Drainage of the wound cavity must be avoided by all means.

Dorsal stabilization of the craniocervical junction will be necessary in many cases because of instability; the utmost care must be taken in the preliminary assessment in order to avoid disastrous consequences (Dastur et al. 1965; Bartal and Heilbronn 1970; Di Lorenzo 1982).

Summary: Transoral Approach

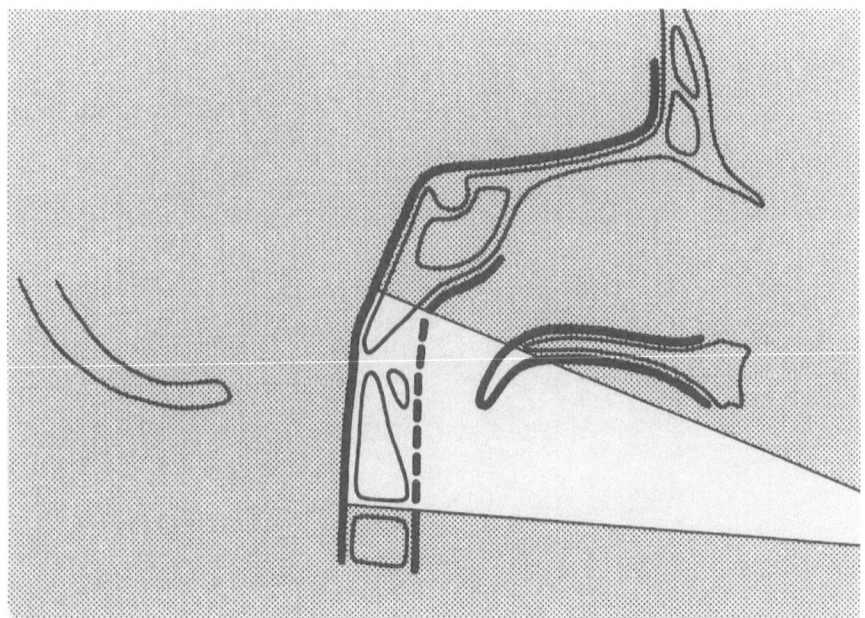

Indications: Extradural tumors of the lower clivus
                  Lesions of the vertebrae C1 and C2

Craniocervical malformations – Chordomas – (Aneurysms)

Advantages: Technically simple

Disadvantages: Risk of infection
                  Risk of CSF leak
                  Dura suture impossible
                  Tracheostomy (?)
                  Craniocervical instability (?)

Limits: C2–C3 (without splitting the mandible and tongue)
                  Lower third of the clivus
                  Close to midline only

Chapter 2: Transcervical Approach

Historical Survey

Stevenson and coworkers were the first to describe the successful removal of a clival chordoma by means of the transcervical approach in 1966. After the results obtained using the transoral and transpalatal approaches had been disappointing, the authors sought an anterior approach that would not entail the drawback of a transmucous approach, namely the danger of infection. The preliminary experiments of White and Albin (1962) and considerable anatomical research of his own finally enabled Stevenson to develop this approach.

In this connection one must not forget to mention the pioneering efforts of Cloward's anterior approach (1958, 1972) as well as Verbiest's lateral approach to the cervical spine, and his commendable work regarding the ventral stabilization of the cervical spine and the craniocervical junction (1968, 1970, 1973, 1977). In 1979, Cloward and Passarelli reported the successful removal of a large clival chordoma by means of the approach they had developed. In doing so, they critically dissociated themselves from Stevenson's transcervical approach, referring rather to their own preliminary work and research done by Verbiest.

Bartel and Heilbronn (1970) removed a clival chordoma by the transcervical approach. Fox (1967) and Wissinger et al. (1967) described the successful clipping of a basilar artery aneurysm by the same approach.

Anatomy

Anatomy Relevant to Surgery

Dissection in the carotid trigone mainly involves the dissection of fasciae and connective tissues in the lateral cervical region. It starts in the region of the middle cervical fascia, in the neurovascular connective tissue of the internal carotid artery, jugular vein, and vagus nerve, and continues downward into the prevertebral fascia (see Fig. 2.6). At the level of the styloid process, the stylopharyngeal fascia, occasionally reinforced in the form of an aponeurosis, covers the main vessels from laterally.

Figure 2.1 shows the position of the ninth to twelfth cranial nerves in relation to the internal carotid artery.

The *ascending pharyngeal artery* originates at the medial circumference of the external carotid artery near the bifurcation of the carotid artery and passes forward between the prevertebral muscles and the posterior wall of the pharynx medial to the internal carotid artery.

The *sympathetic trunk* is situated in the prevertebral fascia on the longus capitis muscle in front of the transverse processes of the cervical spine and adjoins the internal carotid artery near the carotid canal from dorsomedially. It has numerous connections to the vagus nerve.

The branches of the *external carotid artery*, passing forward ventrally, form sufficient collaterals with the contralateral vessels to allow their ligature and transection.

At the intermediate tendon the *digastric muscle* is attached to the hyoid bone, which is where it may be detached and transected.

The *styloid process* serves as the origin of the stylohyoid muscle, whose cloven tendon attaches the digastric muscle to the hyoid bone. The styloid process is also the origin of the styloglossus muscle, which passes forward to the tongue, as well as of the stylopharyngeal muscle. The latter is the guiding muscle of the *glossopharyngeal nerve*, which first runs along the lower side of the muscle and then crosses it laterally in its course to the pharynx. The stylopharyngeal fascia, which is often reinforced by aponeurotic fibres, subdivides the parapharyngeal space and covers the internal carotid artery and the internal jugular vein frontally and laterally (see Fig. 2.6).

Of the caudal cranial nerves, the *hypoglossal nerve* exits at the most medial point; its exit from the hypoglossal canal is situated just dorsomedial to the pars nervosa of the jugular foramen. In its course, the hypoglossal nerve circles the vagus and the internal carotid artery from laterally and passes between the jugular vein and the vagus nerve.

Figure 2.1 depicts the relationships of the cranial nerves and the large vessels in the cervical region. Cervical motor fibres which adjoin the hypoglossal nerve for a short distance have been disregarded in the diagram. They originate at the roots of C2 and C3 and innervate the subhyoid muscles. The caudal cranial nerves in the region of the jugular foramen are located as follows:

The *accessory nerve* exits from the jugular foramen at the most dorsal point and immediately gives off the internal branch to the vagus nerve. In two-thirds of all cases examined, the accessory nerve runs dorsal to the jugular vein, in the remaining third ventral.

The superior ganglion of the *vagus nerve* is situated in the jugular foramen. The internal branch of the accessory nerve and the vagus nerve meet in the area of the inferior ganglion, which is outside the jugular foramen. Past the inferior ganglion the superior laryngeal nerve branches off the vagus nerve and passes forward to the larynx in a slight curve medial to all vessels. In its distal segment, it runs parallel to the superior thyroid artery, where it is easy to identify.

The *glossopharyngeal nerve* leaves the jugular foramen ventral to the vagus nerve and passes forward laterally to the internal carotid artery between the internal carotid artery and the jugular vein. In its further course, it runs along the lower side of its guiding muscle, the stylopharyngeal muscle, circles it from laterally, and proceeds to the pharynx and the tongue.

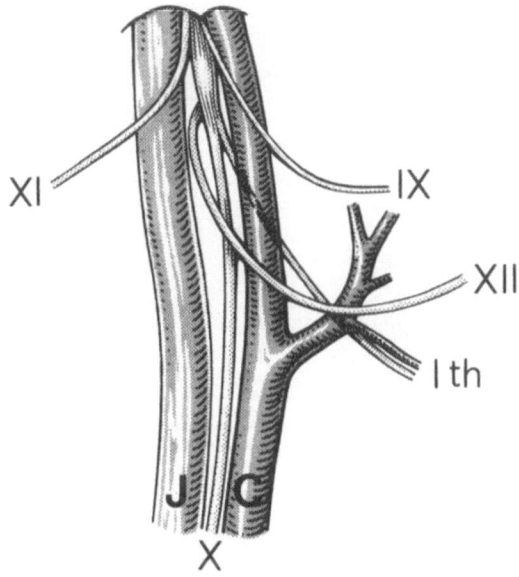

Fig. 2.1. The caudal cranial nerves in relation to the large vessels. *C*, Common carotid artery; *V*, jugular vein; *I th*, superior laryngeal nerve parallel to the superior thyroid artery, the first branch of the external carotid artery; *IX*, glossopharyngeal nerve; *X*, vagus nerve; *XI*, accessory nerve

Key Steps of the Approach

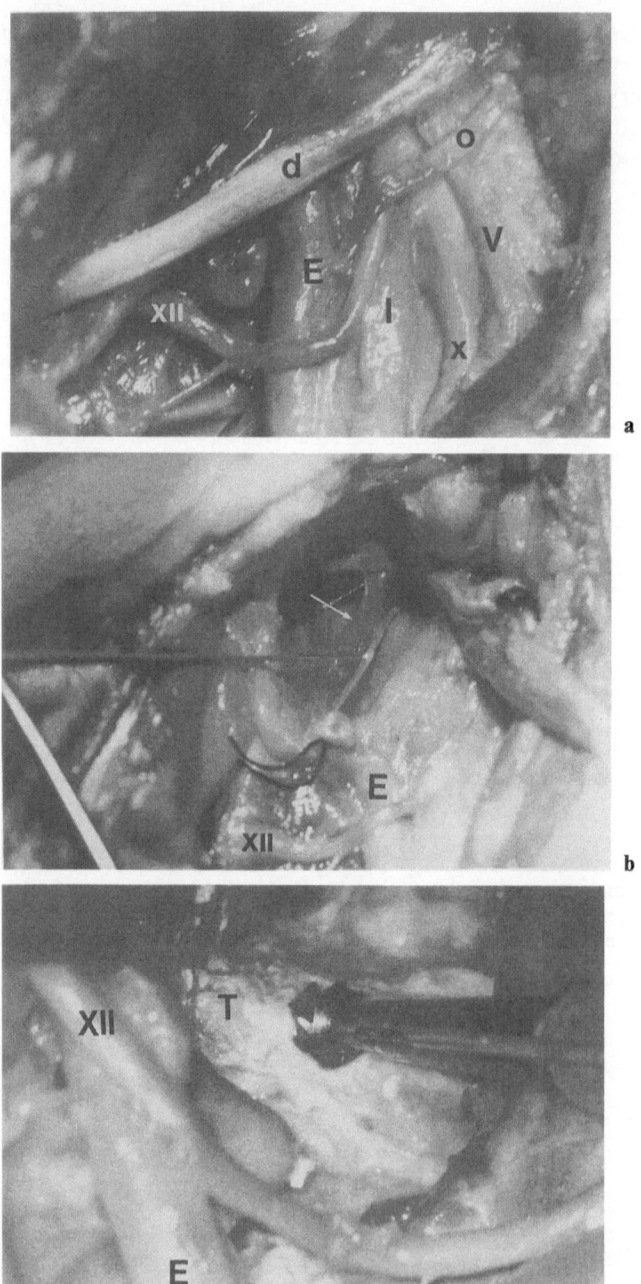

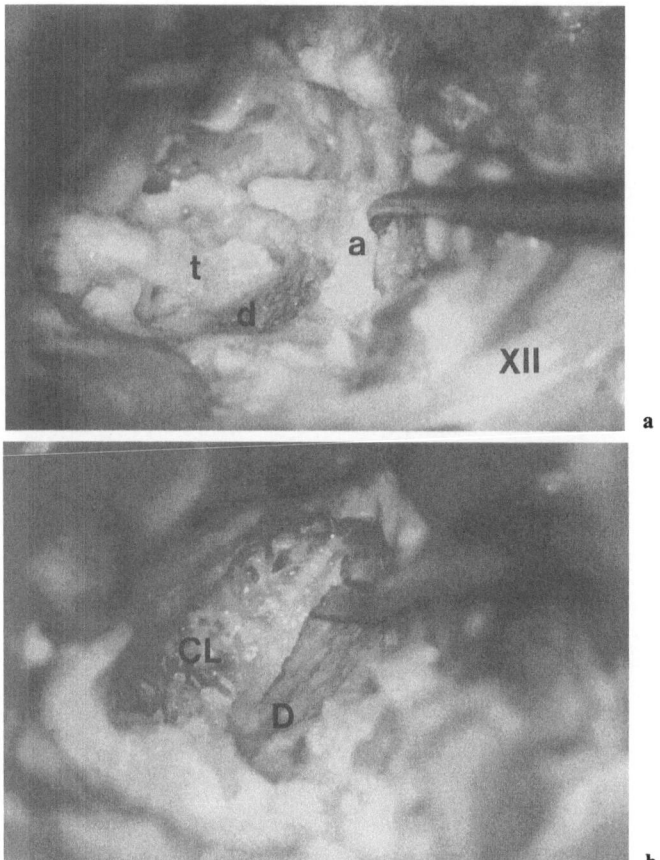

Fig. 2.3 a, b. Osseous resection of the craniocervical junction (**a**) and the clivus (**b**). **a** After the anterior arch of the atlas has been drilled off, the odontoid process is also resected down to its base. Thus the tendons of the craniocervical junction come into view. *d*, Base of the odontoid process; *a*, alar ligaments retracted with dissector; *t*, parts of the transverse ligament; *XII*, hypoglossal nerve. **b** Resection of the clivus (*CL*), a cancellous bone with a marked interior cortical layer. *D*, Clival dura

Fig. 2.2 a–c. Approach in the lateral cervical region: dissection at the carotid trigone from the left (**a**), when reaching the retropharyngeal space (**b**), and beginning of resection of the anterior arch of the atlas (**c**). **a** Digastric muscle (*d*) with intermediate tendon, hypoglossal nerve (*XII*) and descending branch. *X*, vagus nerve; *E*, external carotid artery; *I*, internal carotid artery; *V*, jugular vein; *o*, occipital artery. **b** After transection of the occipital artery and the digastric muscle: view of the glossopharyngeal nerve (*hook*) and stylopharyngeal fascia (*white arrows*). **c** Beginning of resection at the arch of the atlas with a drill. *T*, Anterior tubercle of the atlas

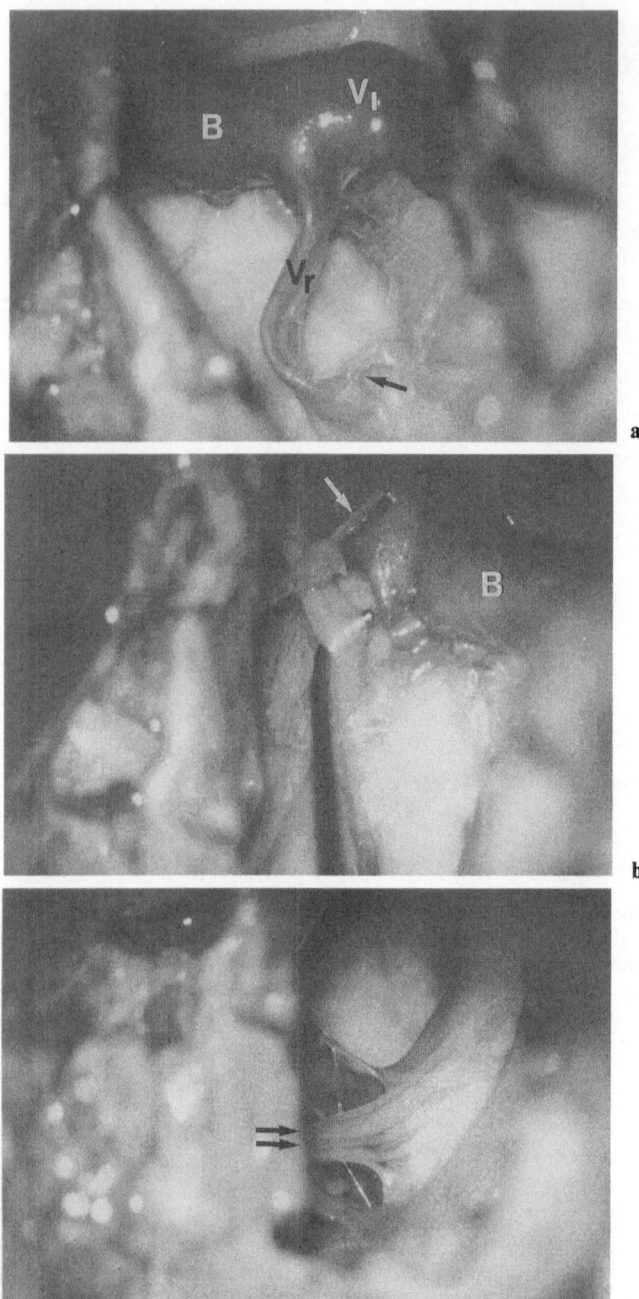

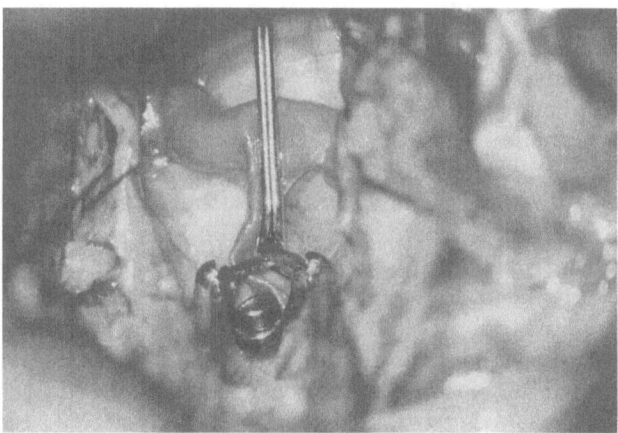

Fig. 2.5. Comparison of dimensions of clival craniotomy to an angular clip 1.2 cm long

Surgery

Technique

The patient's head is fixed in reclination and turned to the opposite side; the skin incision leads from the tip of the mastoid to the tip of the chin parallel to the mandible. It is possible, though not mandatory, to enlarge it laterally in the shape of a T over the sternocleidomastoid muscle. The carotid trigone is dissected, the internal carotid artery is snared, and the superior thyroid and lingual arteries are ligated at their origins and transected. The hypoglossal nerve is largely dissected, but not severed (as was described by e.g., Fox in 1967). The internal carotid artery, jugular vein, and vagus nerve are exposed as far as the osseous skull base. Dissection of the submandibular trigone facilitates the exposure and mobilization of larynx and pharynx. The retropharyngeal space is bluntly dissected. The superior constrictors of the pharynx must be detached from the pharyngeal tubercle. The compact, often aponeurotic, stylopharyngeal fascia covers the large vessels from frontally and laterally; it is also transected. If necessary, the styloid process is resected, which results in an extension into the parapharyngeal space. The internal carotid artery, jugular vein, and ninth, tenth, and eleventh cranial nerves can be

---

Fig. 2.4 a–c. After incision of the dura, the vertebral junction (**a**) as well as the cranial (**b**) and contralateral boundaries (**c**) of the approach are exposed. *B* Basilar artery; V_l, left vertebral artery; V_r, hypoplastic right vertebral artery; *black arrow*, origin of anterior spinal artery. Right abducent nerve exposed with a nerve hook; *white arrow*, anterior inferior cerebellar artery; *double arrow*, root fibres of the hypoglossal nerve

displaced laterally. The nasopharynx is subperiostally detached from the clivus until the vomer is palpable.

After the prevertebral fasciae have been split along the median from the clivus to the level of C3, the prevertebral muscles are subperiostally removed as far as the transverse processes. Thus, the limits of operation at the exterior aspect of the skull base can be described as follows: laterally from the place where the internal carotid artery enters the carotid canal, frontally to where the vomer is palpable (the pharyngeal tubercle is situated at a distance of about 11 mm from the anterior margin of the foramen magnum), and again laterally as far as the pterygoid processes permit.

After the anterior arch of the atlas and the odontoid process have been drilled off, the osseous opening measures about 25 × 35 mm. The boundaries of the operation site on the interior aspect of the skull base are represented by the internal carotid artery, inferior petrosal sinuses, jugular tubercles (be careful of the hypoglossal nerve) and the spheno-occipital suture (see Fig. 2.7). After successful tumor excision the osseous defect is packed with muscle and the wound is closed in multiple layers. In case of dural excision a dural graft (a patch of lyophilized dura or fascia lata and muscle) is fit in and sealed with fibrin glue. In addition, we institute external lumbar drainage for some days.

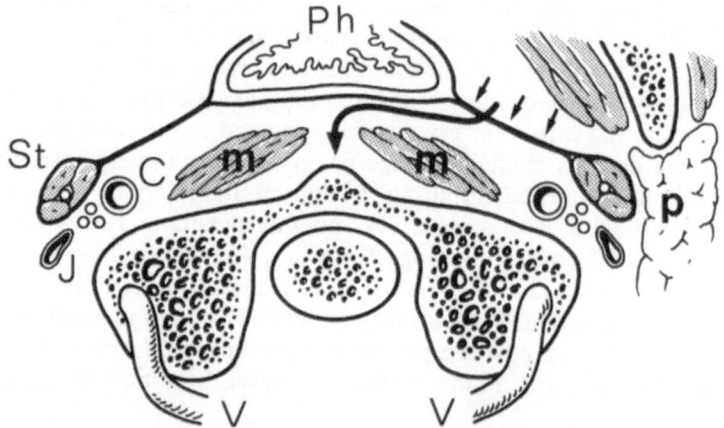

Fig. 2.6. Horizontal section at the level of C1. The clivus is approached by dissection in the parapharyngeal and retropharyngeal spaces (*braced arrow*). The *arrow* points straight at the anterior tubercle of the atlas. The stylopharyngeal fascia (*small arrows*) has to be split. *Ph,* Pharynx; *m,* prevertebral muscles; *p,* parotid gland; *C,* internal carotid artery; *J,* jugular vein, with ninth, tenth, and eleventh cranial nerves between; *V,* vertebral artery; *St,* styloid process and attached muscles

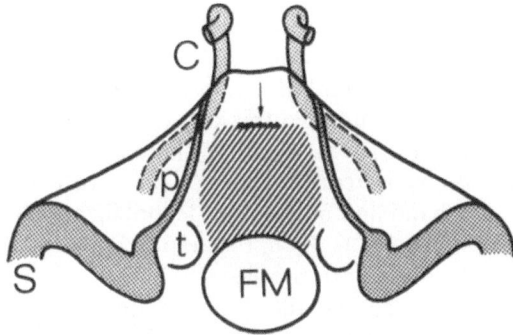

Fig. 2.7. Intradural aspect of the clival fenestration following transcervical approach. Clival craniotomy is represented by the *hatched area*. Note that the intradural structures of the contralateral side are more easily accessible than those of the ipsilateral side. *Arrow*, Spheno-occipital suture; *C*, internal carotid artery; *p*, inferior petrosal sinus; *FM*, foramen magnum; *t*, jugular tubercle *S*, sigmoid sinus

Frequently Encountered Lesions

Since this approach represents a further development of the anterior approach to the cervical spine, the main indications for its application are lesions encroaching on the clivus from the upper cervical spine or on the cervical spine from the clivus. These may be *vertebral metastases* (Nagashima et al. 1979), *chordomas* or *plasmocytomas*, or *traumatic changes* of the craniocervical junction (Lesoin et al. 1986; Böhler 1982). In cases in which vertebral bodies and articular processes are affected by the tumor, the vertebral artery may be a problem during surgery. The artery may pass forward far laterally, especially between C1 and C2, where it may easily be damaged.

When using the transcervical approach in clipping *basilar aneurysms*, the anatomical situation will usually be found to be normal. It must, however, be borne in mind that the approach is slightly extra-axial, i.e., the view of the contralateral side in the depth of the operation area is better.

Advantages and Disadvantages of the Approach

The essential advantage of the transcervical approach over the transoral approach lies in the purely submucous dissection, which results in a considerable reduction of the risk of infection. In case of CSF leak, there is no communication with the oral cavity; its management may, however, still be difficult (Fox 1967). Moreover, the surgeon may fall back on methylacrylate to replace vertebrae (Lesoin et al. 1986; Nagashima et al. 1979).

For the reasons mentioned above, the transcervical approach is not only limited to extradural lesions, provided that they are located in the lower half of the clivus. Of course, the limitations only pertain to the clival extent: the entire cervical spine can be attained this way (Lesoin et al. 1986). The site of operation is laterally restricted, which is characteristic of all anterior approaches. Tracheostomy is recommended to achieve complete mandibular occlusion on the one hand and not to impede mobilization of the pharynx on the other hand (Stevenson et al. 1966; Fox 1967). We think that nasal intubation with a soft tube solves both problems.

As the arch of the atlas and the odontoid process have been removed, special attention must be paid to a potential development of craniocervical instability (De Andrade and MacNab 1969). In any given case, the necessity of dorsal stabilization will have to be decided on. There are reports of fatal complications as well as acute tetraplegia, each occurring during the interval between successful clival surgery and the envisaged dorsal fusion (Bartal and Heilbronn 1970; Di Lorenzo 1982). The depth of the operation site should not represent a major problem when an operating microscope is used.

So far, we have not encountered any case among our patients that provided a cogent indication for the use of this approach. One reason is that we have quite successfully employed the transoral approach (see Chap. 1), which technically speaking is somewhat simpler. Moreover, we do not consider the transcervical approach fully adequate for midbasilar aneurysms (see Advantages and Disadvantages sections in Chaps. 1 and 8).

Summary: Transcervical Approach

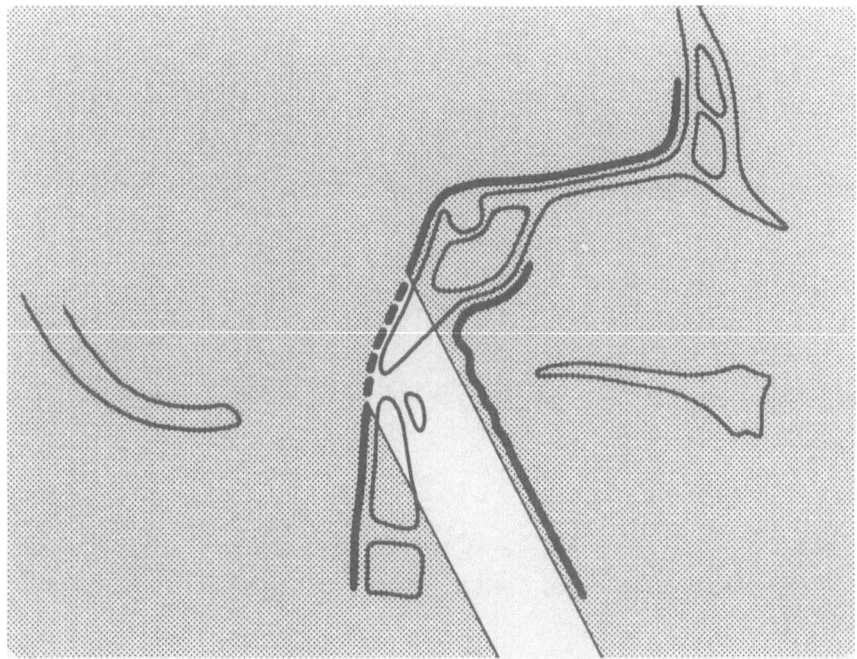

Indications: Lesions of the lower clivus and the upper cervical spine
which may extend intradurally

Chordomas – Metastases – Aneurysms

Advantages: Transdural surgery possible
No increase in the risk of infection
Synthetic materials can be used to replace vertebrae

Disadvantages: Deep field of operation
Craniocervical instability (?)
Tracheostomy (?)
Extra-axial approach

Limits: Close to midline only
Spheno-occipital suture
Inferior petrosal sinus
Internal carotid artery

Chapter 3: Transsphenoidal Approaches

Historical Survey

At the beginning of our century, in 1906, Schloffer was the first to describe an extradural approach to the pituitary gland (Schloffer 1906, 1907), and thus became a pioneer of numerous similar approaches with the common goal of removing pituitary adenomas without craniotomy, for in those days subfrontal craniotomy caused much greater strain and risk for the patient than now. Following that innovative step, as is often the case, a variety of similar extradural approaches to the pituitary gland were developed. Most interesting among them and most frequently used up to today are the endonasal approach as described by Hirsch (1909, 1910, 1958) and the transethmoidal approach of Chiari (1912). The transpalatal approach described by Tiefenthal in 1920 and the transmaxillary approach of Fein (1910), later on modified by Denker (1921) and Lautenschläger (1929), are no longer in commonly use. Archer et al. (1987) described a transclival approach after bilateral maxillotomy, to clip aneurysms of the vertebrobasilar system. The approach most frequently employed in neurosurgery today, the sublabial-transseptal-transsphenoidal approach, was developed by Cushing as early as 1912 and combines optimal cosmetic results with strictly midline dissection. Guiot, who championed this technique, continued its development and performed surgery under the microscope (Guiot 1958, Guiot and Thibaut 1958; Guiot and Derome 1976).

In 1965, Hardy, who studied with Guiot, introduced lateral intraoperative fluoroscopy (Hardy and Wigser 1965), which has been in routine use ever since (Hardy 1965, 1969 b, 1971). Subsequently, the transseptal-transsphenoidal approach was used not only for endo- and suprasellar pituitary adenomas, but also for lesions extending from the clivus (Spiess 1911; Guiot 1968; Rougerie et al. 1967; Decker and Malis 1970; Hardy 1977; Derome and Guiot 1979). ENT surgeons were particularly involved in extending the transethmoidal variation of the approach to the pituitary gland; after unilateral or bilateral resection of the ethmoid along the osseous skull base, the clivus can be reached via the sphenoidal sinus.

Anatomy

Anatomy Relevant to Surgery

Transseptal-Transsphenoidal Approach

Midline orientation is of paramount importance for the transseptal approach. The median is marked by two points of orientation: the anterior nasal spine and the median rostrum of the sphenoid. The *nasal septum*, which is located between these two points, is usually not situated along the median and cannot be used for midline orientation, as is also the case with the septa of the sphenoidal sinus.

The mucous membrane of the nasal septum is abundantly supplied with blood from branches of the sphenopalatine, anterior, and posterior ethmoidal arteries from dorsally and cranially as well as from a branch of the superior labial artery in the anterior portion of the septum.

The distances relevant to the transseptal-transsphenoidal approach are shown in Fig. 3.1.

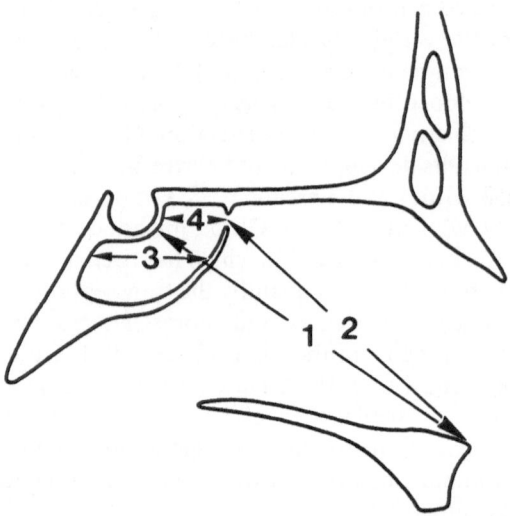

Fig. 3.1. Distances in transsphenoidal approaches

1 Anterior nasal spine to sellar floor   72 mm (Dixon 1937, cited in Lang (1981)
2 Anterior nasal spine to sphenoidal sinus   59 mm (Lang 1981)
3 Length of sphenoidal sinus   24 mm (Lang 1981)
4 Anterior wall of sphenoidal sinus to sella   17 mm (Fujii et al. 1979)

Osseous dimensions of the sella

Depth	8 mm	(4–12 mm)	(Camp 1924, cited in Lang 1981)
Length	10.5 mm	(5–16 mm)	(Camp 1924, cited in Lang 1981)
Width	14 mm	(10–16 mm)	(Renn and Rhoton 1975)

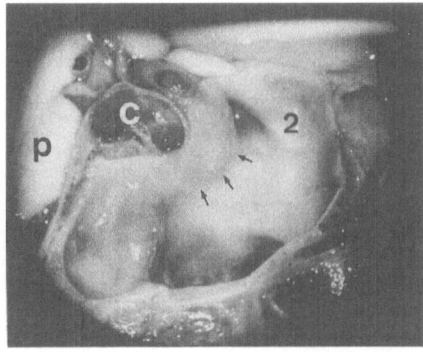

Fig. 3.2. Paramedian sagittal section of the sphenoidal sinus: view of the lateral wall of a well-pneumatized sinus (sellar type) on the left, with a markedly developed carotid prominence (*arrows*) and protrusion of the optic canal (*2*). In between is the optico-carotid recess. *p,* pons; *c,* section of the internal carotid artery

The *sphenoidal sinus* already exists at birth in the form of an evagination of mucous membrane into the body of the sphenoid bone: however, it develops dorsally during puberty.

According to Hamberger, a well-pneumatized sphenoidal sinus (sellar type) is found in 86% of adults, in 11% the sinus is pneumatized in the anterior portion only (presellar type), whereas in 3% a conchal type with very little pneumatization is found (Hamberger 1961, cited in Lang 1981, p. 152). The sphenoidal sinuses often develop asymmetrically, with slanted, rudimentary, or even transverse septa. The lateral wall of the sphenoidal sinus is important for its immediate neighborhood. In cases of good pneumatization, the anterior knee of the internal carotid artery (carotid prominence) and the optic nerve protrude into the sphenoidal sinus (Fig. 3.2). There is evidence for this in 70% of cases according to Renn and Rhoton 1975 and up to 96% according to Fujii et al. 1979). In fewer cases, and depending on the pneumatization of the sphenoidal sinus, the maxillary nerve – in even rarer instances the mandibular nerve or the pterygoid canal – may protrude into the sinus in the area of the lateral wall (Van Alyea 1941; Fujii et al. 1979; Lange 1981). The bone covering these structures is usually very thin and may even be lacking (Renn and Rhonton 1975).

According to Fujii et al. (1979), the average *distance between the carotid arteries* is 17 mm (8 – 24 mm). They may, however, also be located at a distance of 4 mm from each other (Renn and Rhoton 1975), or may even touch (Weiss 1987).

Transethmoidal-Transsphenoidal Approach

The anatomy pertaining to the transethmoidal-transsphenoidal approach will be described in Chap. 4.

Key Steps of the Approach

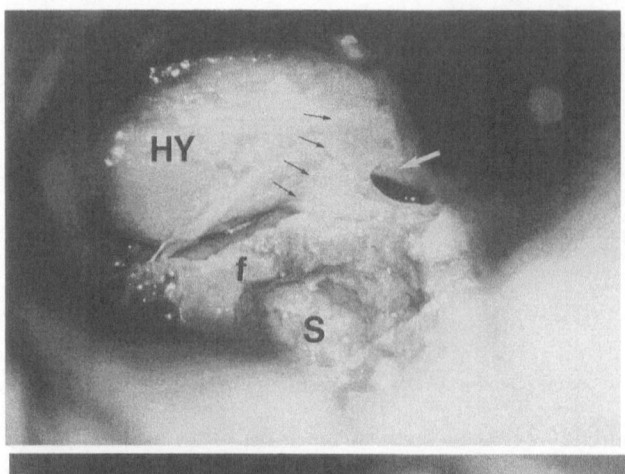

a

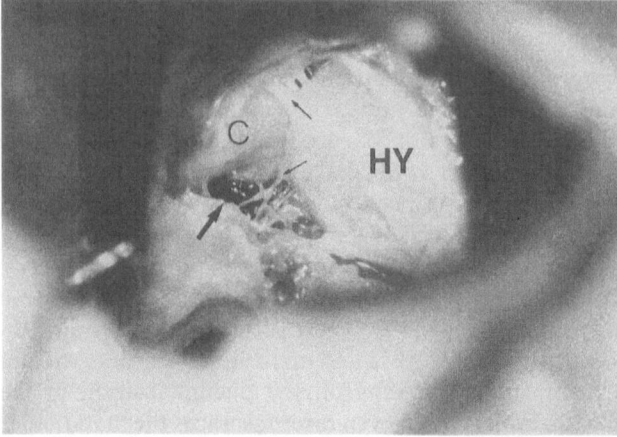

b

Fig. 3.3. a After sublabial-transseptal-transsphenoidal approach, the anterior wall of the sella turcica is removed and the pituitary gland exposed (*HY*). The tender periosteal lining of the sella floor adjoins the boundary of the cavernous sinus on the left (*small arrows*). When it is opened the intracavernous segment of the left internal carotid artery is exposed (*white arrow*). *f*, Sellar floor; *S*, sphenoidal sinus. **b** The cavernous sinus on the right side (*large arrow*) shows so-called trabeculae (*small arrows*) and the intracavernous segment of the internal carotid artery (*C*). *HY*, pituitary gland

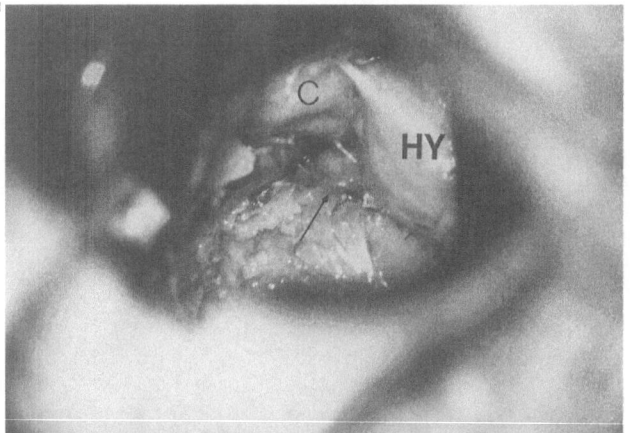

a

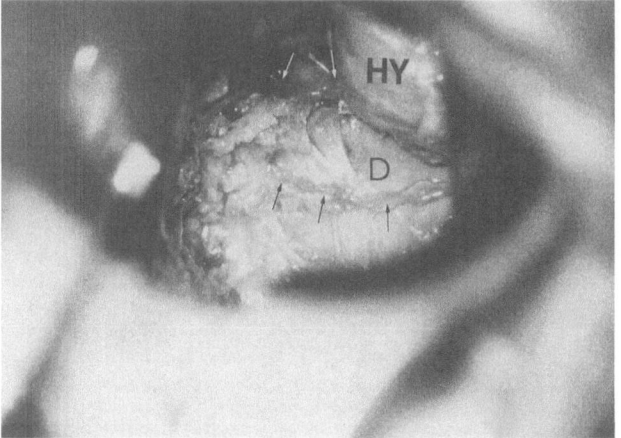

b

Fig. 3.4. a Further preparation of the cavernous sinus exposes the horizontal segment of the intracavernous internal carotid artery (*C*) and the inferior hypophyseal artery (*arrow*), which originates from the meningohypophyseal trunk. The artey enters the pituitary capsule from laterally and passes forward between the anterior and posterior lobes of the pituitary gland. *HY*, pituitary gland. **b** Further dissection exposes the clival dura (*D*) and the medial clival artery (*arrows*), as a branch of the meningohypophyseal trunk and the inferior hypophyseal artery (*white arrows*)

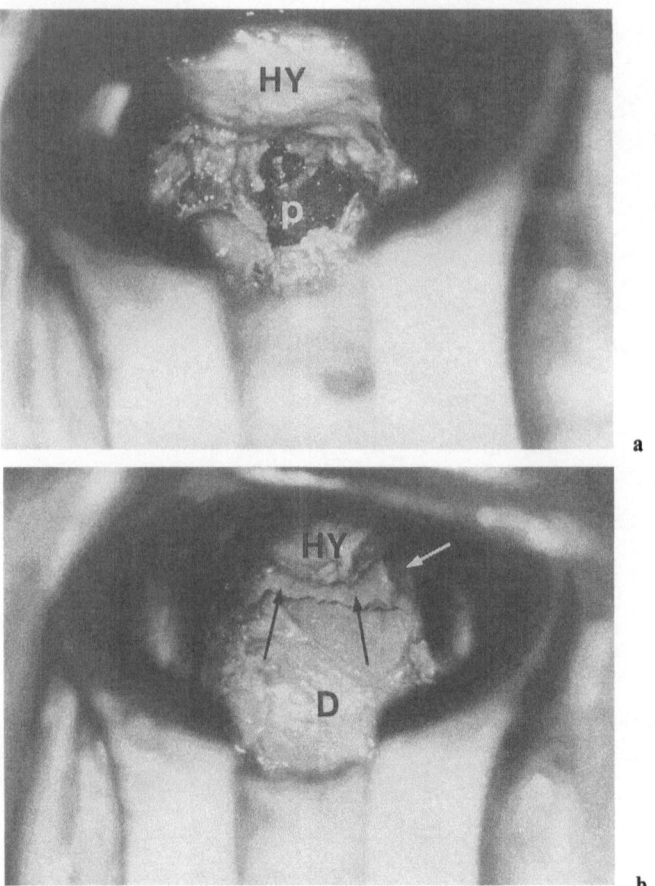

a

b

Fig. 3.5. **a** Before reaching the clival dura a marked venous basilar plexus is found (*p*). **b** The dorsum sellae (*black arrows*) represents the nonaccessible part of the osseous clivus. See also Fig. 3.9b. *HY*, Pituitary gland; *D*, clival dura; *white arrow*, portion of the dorsal knee of the internal carotid artery on the left adjoining the dorsum sellae

Surgery

Technique

Transseptal-Transsphenoidal Approach

The patient is placed in a supine position with his head turned toward the surgeon, who stands at his side facing him. An alternative position would be to leave the patient's head straight and fix it in slight reclination; in this case, the surgeon stands at the head. Fluoroscopy is centered on the clivus from exactly laterally so the direction of approach and the positions of the instruments can be checked (Fig. 3.7).

In the *sublabial modification* of this approach the mucous membrane of the nasal septum is infiltrated first and a sublabial incision performed. The lower limit of the nasal aperture is exposed. The mucous membrane is detached unilaterally from the cartilaginous septum, which is then transected at the transitional zone to the osseous septum and mobilized. Thus, the cartilaginous part of the septum is displaced along with the contralateral mucous membrane like the wings of a double door. The mucous membrane ist detached bilaterally along the osseous septum only, and the latter is resected in a piecemeal manner until the anterior aspect of the sphenoid sinus is reached.

In the *endonasal modification*, the incision of the mucous membrane is placed several millimeters behind the frontal rim of the septum and reaches to the floor of the nose. Further submucous dissection is carried out as outlined above.

Fluoroscopic monitoring of the desired direction of approach is begun during resection of the osseous part of the septum. Removal of the anterior aspect of the sphenoid bone often already exposes the tumor boundary. Sufficient pneumatization of the sphenoidal sinus greatly facilitates access to the clivus. First, endocapsular tumor resection is carried out with the help of bipolar coagulation, suction, and various curettes; the tumor is reduced in size and excised until the osseous boundary is attained. For demarcation of the tumor from the basal dura, sound dura must be exposed all around. Whereas in purely epidural tumors the clival dura forms a clearly discernible boundary, tumors invading the dura necessitate the exposure of unaffected parts of the dura to make resection possible.

This is the only feasible way to complete removal of a tumor from the brain stem structures. Dural defects, if any, are covered with lyophilized dura, fascia lata, and/or muscles and sealed off with fibrin glue. In such cases final tamponage is of paramount importance.

We additionally institute external lumbar drainage for a few days postoperatively to prevent CSF leakage.

Transethmoidal-Transsphenoidal Approach

The skin incision is begun at the inner canthus and is either directed toward the eyebrow or continued downward paranasally, depending on tumor size and site of origin. After exposure of dacryocyst and trochlea, the periorbit and its contents are displaced and the inferior wall of the frontal sinus is opened. The anterior wall of the sphenoidal sinus is reached as the ethmoid is removed in a piecemeal manner along the anterior skull base. Fenestration of the anterior sphenoidal wall makes further dissection toward the sella and clivus possible.

In this way the optic canal and the anterior knee of the internal carotid artery can be exposed. With due attention to the internal carotid artery, the tumor may now be removed caudally and medially until, step by step, the clival dura is reached. As the approach is para-axial, the contralateral internal carotid artery must be taken into account in the depth of the operation area. This procedure results in wide communication of tumor bed and nasal cavity. Therefore, the resection cavity must be packed carefully. If dura has to be resected, lyophilized dura, fascia lata, muscle, or a piece of galea aponeurotica originating from the superciliary incision is fit onto the defect and sealed off with fibrin glue. The dural graft is supported by nasal packing. In such cases, the packing is left for a longer period than in solely extradural excisions; in addition, external lumbar drainage is instituted to prevent CSF leakage. Dimished CSF pressure and the counterpressure created by the tamponage enable the dural graft to settle quite rapidly and greatly reduce the risk of persistent CSF leakage.

Transfacial Approach

The transfacial approach (Jackson et al. 1986) represents a further development of anterior approaches to the skullbase with the skin incisions along facial folds. These cosmetically unsatisfactory incisions are justified in cases of malignant tumors of the viscerocranium, where the resection has to include parts of the facial skeleton and overlying skin. One possible advantage of face-splitting approaches (Janecka and Shekar 1988; de Fries et al. 1988) is the possibility of "en bloc" resections. After maxillotomy a wide approach results which combines the advantages of a transethmoidal, transsphenoidal, and transoral approach. This was for us the reason to use this approach for the removal of an exceptionally large recurrent clival chordoma. Otherwise, in tumors of the skull base this approach is rarely needed.

Case Report (Transsphenoidal Approach)

This 45-year-old woman complained of headache; the only neurological deficit was paresis of the right abducent nerve. There was no evidence of endocrinological disorders. CT revealed an inhomogeneous sellar and subsellar lesion without suprasellar extension. The clival dura appeared intact and extended only slightly toward the brain stem (Fig. 3.6b). As the lesion extended to the lower third of the clivus and was in all probability extradural, the transseptal-transsphenoidal approach was well suited well to it. The operation was performed using the technique outlined above (Fig. 3.7). As the process was a fibrous chordoma, extirpation was rather difficult; the tumor had to be excised in a piecemeal manner. It was a purely extradural tumor; the compact dura of the clivus formed a good boundary. In the lower third of the clivus resection became increasingly difficult. Macroscopic radical tumor extirpation was carried out (Fig. 3.8).

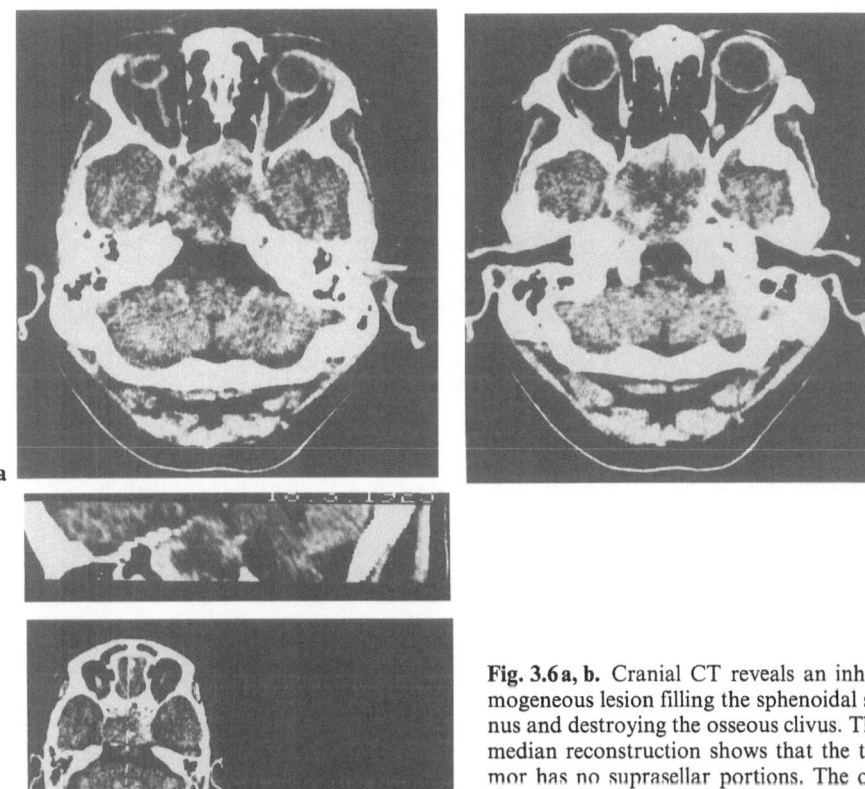

Fig. 3.6a, b. Cranial CT reveals an inhomogeneous lesion filling the sphenoidal sinus and destroying the osseous clivus. The median reconstruction shows that the tumor has no suprasellar portions. The osseous parts of the dorsum sellae and of the upper two-thirds of the clivus have been destroyed

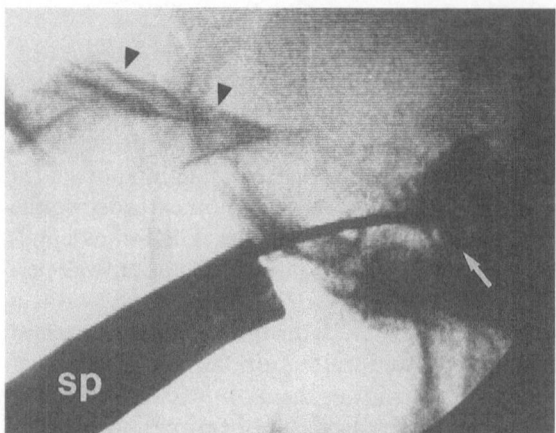

Fig. 3.7. Lateral intraoperative fluoroscopy. The speculum (*SP*) reaches the anterior wall of the sphenoidal sinus; the contours of the clivus are not visible. The *white arrow* shows the tip of an instrument at the lower third of the clivus

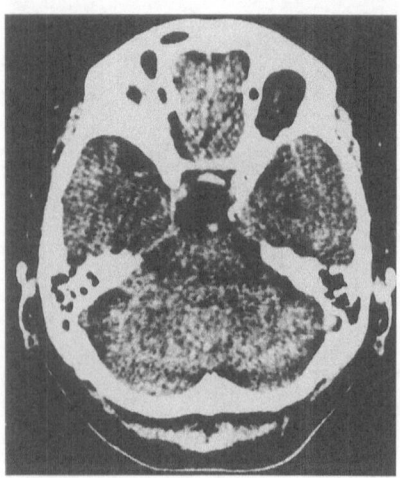

Fig. 3.8. CT after tumor extirpation

Fig. 3.9 a–c. Axial (**a**) and coronal (**b, c**) CT scans show the amount of bony destruction of the clivus, with the tumor bulging into the nasopharynx. The apex of the petrous bone has been eroded on both sides (**c**)

Case Report (Transfacial Approach)

This 53-year-old lady (Figs. 3.9–3.14) suffered from a huge recurrent chordoma of the clivus, which extended from the dorsum sellae to the odontoid process. The tumor extended far laterally into both petrous bones, resulting in hearing loss on the right side. In addition to this there was incomplete palsy of the sixth and complete palsy of the seventh, ninth, tenth, and twelfth cranial nerves on the right side. She also had pronounced hemiparesis of the left side and left cerebellar signs. CT and MRI revealed a huge tumor recurrence involving the whole clivus with lateral extensions into both petrous apices. On angiography the basilar artery was shown to be markedly displaced; the intracavernous part of the internal carotid artery was also displaced by tumor (Fig. 3.11).

After tracheostomy a transfacial approach was performed with the skin incision as outlined in Fig. 3.12a. The Le Fort osteotomy was performed with median transection of the soft palate, giving a wide exposure from the sphenoid bone to the atlas (Fig. 3.12b). The tumor bulged into the pharynx and was covered by mucosa only. After median pharyngeal incision the chordoma was resected piecemeal. The bone of the clivus was completely destroyed with normal dura just at the level of the posterior clinoid processes. After tumor removal the basilar artery and the pons could be visualized (Fig. 3.13).

Due to the large dural defect, closure was performed with lyophilized dura and fibrin glue. Unfortunately it was not possible to achieve safe, multilayered pharyngeal closure. The osteotomies were fixed by miniplates.[1]

The postoperative course was complicated by a CSF leak with consequent meningitis, requiring prolonged postoperative intensive care and finally a permanent lumboperitoneal shunt.

---

[1] We would like to thank Prof. Hausamen of the Department of Maxillofacial Surgery, Hannover, for his help and cooperation during surgery.

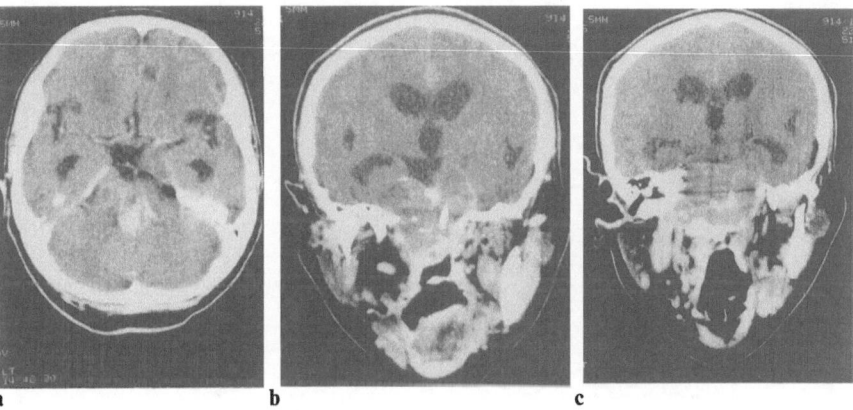

a   b   c

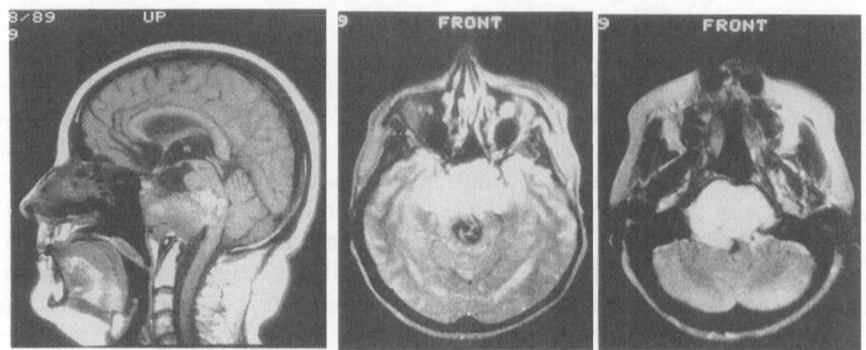

a, b   c

Fig. 3.10a–c. MRI shows a huge chordoma of the clivus extending from the dorsum sellae
to the foramen magnum (**a**). Laterally the tumor extends into both temporal fossae. The
cavernous sinus has been invaded on both sides and the sphenoid bone destroyed. The brain
stem is severely compressed, worst of all at the level of the lower pons

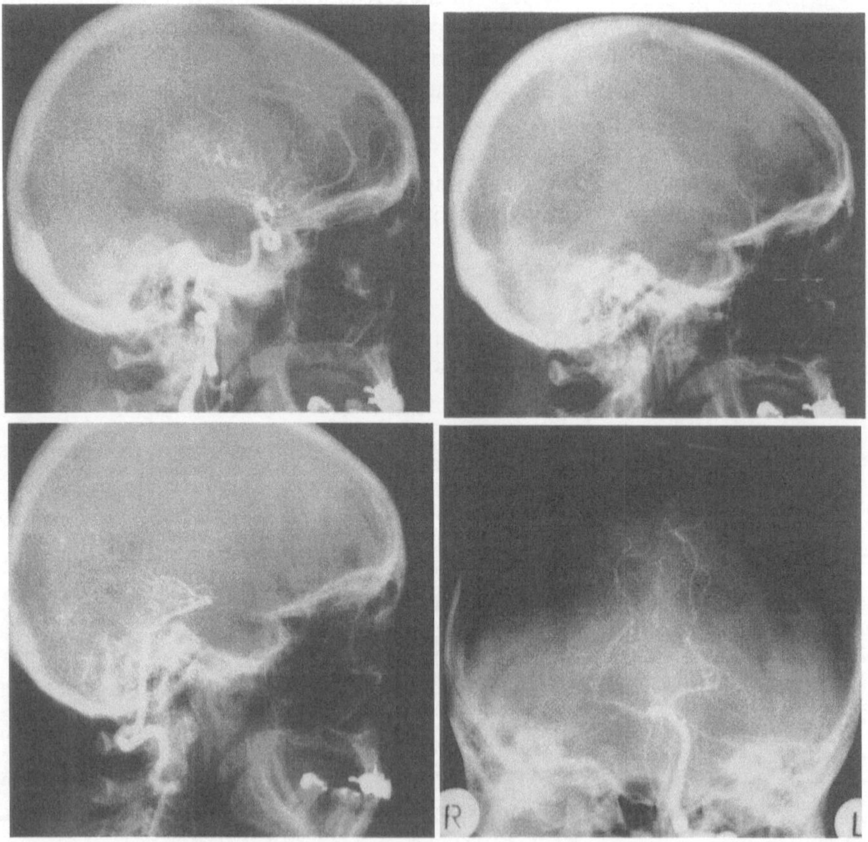

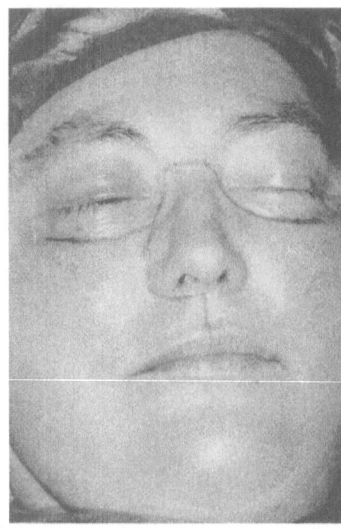

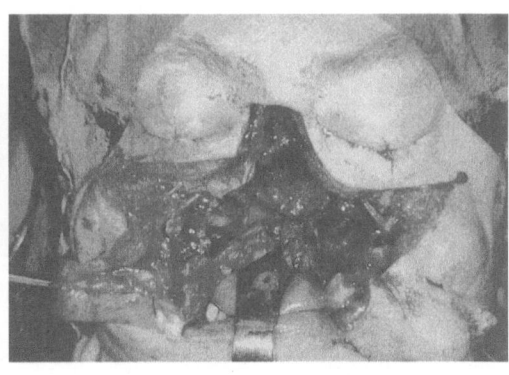

b

Fig. 3.12. a The patient with the skin incision outlined preoperatively. **b** The osteotomy of the facial
skeleton provides a broad approach from the skull
base of the atlas

a

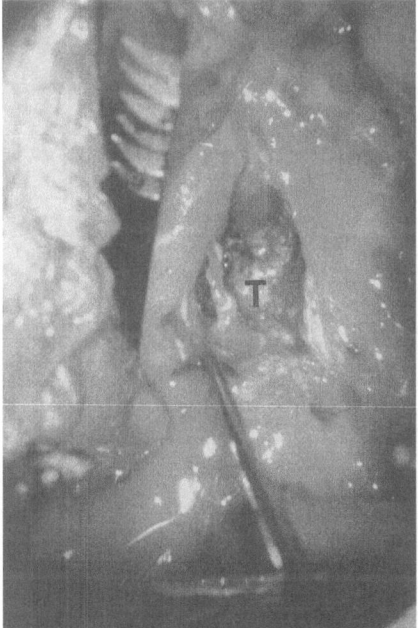

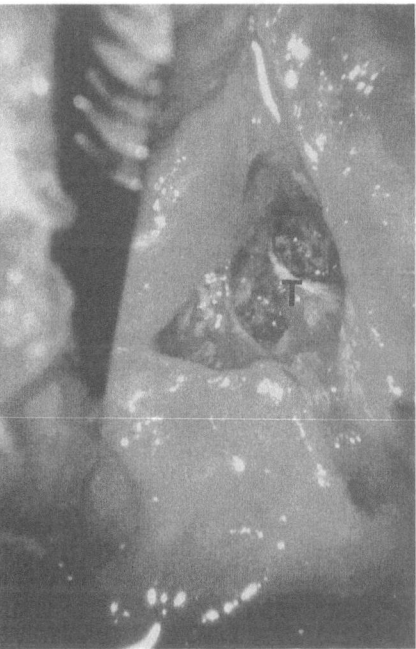

a

b

Fig. 3.13 a, b (Legend see page 46)

◄ ───

Fig. 3.11 a–d. Angiography of the right carotid artery (**a, b**) and left vertebral artery (**c, d**)
demonstrates the prepontine mass lesion with almost no tumor staining. The intracavernous
segment of the carotid artery is also displaced by tumor

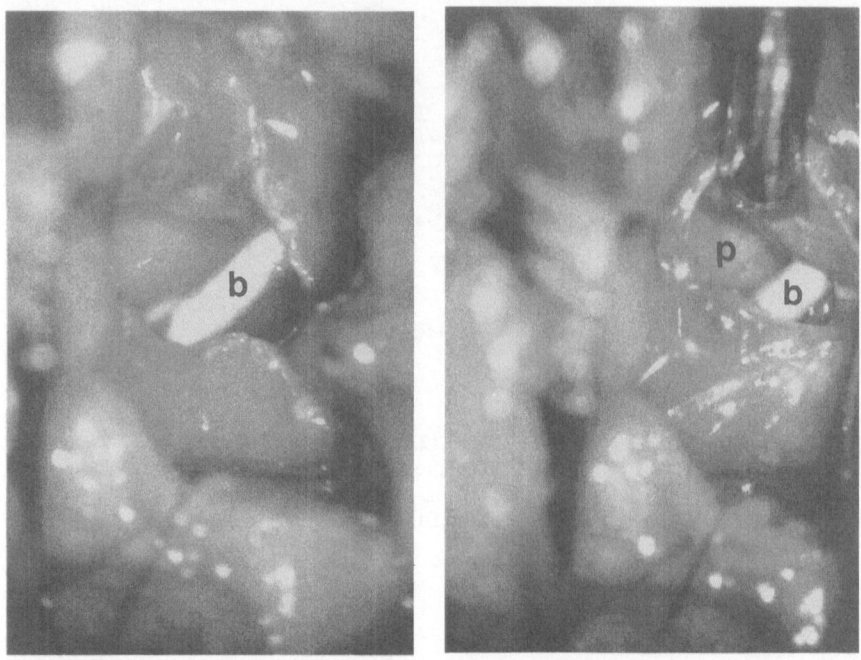

c   d

Fig. 3.13a–d. The pharyngeal wall has been split (**a**) and the underlying tumor is approached (**b**). After tumor removal the basilar artery and pons could be visualized (**c, d**). Note the large defect of the clival dura. *T,* Tumor; *b,* basilar artery; *p,* pons

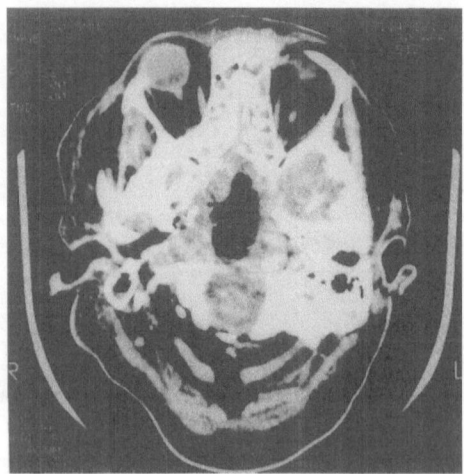

Fig. 3.14. Postoperative CT shows the large defect at the level of the foramen magnum

Frequently Encountered Lesions

The main indications for the transseptal-transsphenoidal approach are *pituitary adenomas* encroaching on the sphenoidal sinus and the adjoining upper clivus. Such pituitary adenomas may invade the dura as well as the bone. Instances in which they do not extend to the suprasellar region are rare, but so long as the suprasellar extension is not polycyclic and does not encase vessels, such tumors, too, can be extirpated by transseptal-transsphenoidal surgery. Mostly, adenomas are of soft consistency, which facilitates removal as they can be resected by suction and curettes. So even they extend to the middle and lower clivus this does not make surgery too difficult.

The situation is different in *chordomas* or *chondromas*, whose consistency is often fibrous and resistant. They destroy the osseous part of the clivus and invade the dura (although only at a more advanced stage). As long as the compact dura of the clivus forms a layer separating the tumor and the brain stem, transseptal or transethmoidal extirpation is possible. If the tumor has reached the lower third of the clivus, resection becomes difficult, and it is often impossible in cases when the tumor extends to the level of the foramen magnum.

Carcinoma of the paranasal sinuses invades the dura at an early stage and may also infiltrate the anterior cranial fossa or extend in the direction of the sphenoidal sinus. As long as important structures such as the internal carotid artery, the optic nerve, or the cavernous sinus can be identified, such tumors may be resected by the transethmoidal approach even if they extend to the clivus.

Advantages and Disadvantages of the Approaches

Transseptal-Transsphenoidal Approach

In our opinion, this is an approach that makes it easy to reach the upper third of the clivus, in particular for surgery of sub- or retrosellar pituitary adenomas. The larger the caudal extension of the lesion, the more difficult its resection. Opinions about the caudal limits to resection on the clivus differ greatly. Reaching the upper third of the clivus ought to be relatively safe and unproblematic as stated by Derome and Guiot (1979). According the Rougerie et al. (1967), the caudal limit is approximately 1 cm above the foramen magnum. For Decker and Malis (1970), the lower limit of what is feasible is at the level of the hard palate, because they think that the clivus slopes away dorsally, for which reason further resection appears too risky. Hardy (1977), by contrast, demonstrated in four cases that reaching the foramen magnum is possible.

The key to the feasibility of tumor excision by the transsphenoidal approach is definitely tumor consistency. This goes particularly for tumors extending laterally. The lateral limits to resection are thus the major drawback of this approach; removal of hard and compact tumors is extremely difficult this way. One of the great advantages, however, is that the approach does not put too much strain on the patient, so it is also suitable for high-risk patients in whom a transcranial operation would be too hazardous and therefore contraindicated. Endocapsular tumor resection usually results in sufficient decompression of the brain stem. In addition, tumor biopsy may be obtained by this approach in cases where stereotactic biopsy is impossible.

Transethmoidal Approach

Although tumors of the sphenoid bone usually force the carotid arteries apart and thus widen the access to the clivus, para-axial dissection must be regarded as disadvantageous. The risk of damaging the ipsilateral internal carotid artery proximally and the contralateral internal carotid artery distally is higher in the transethmoidal than in the transseptal approach. It is extremely difficult to view the portion of the tumor behind the ipsilateral internal carotid artery. This shortcoming can only be made up for bilateral transethmoidal surgery: the incision along the eyebrow has to be extended to the second eyebrow via the glabella. Thus, the operation site is enlarged between the carotid arteries toward the clivus along the osseous anterior skull base. This procedure also ensures midline orientation. In general, the anterior skull base dura is the cranial limit (Fig. 3.15 b). In tumors of pre- or suprasellar extension the basal dura would have to be opened and/or resected. In such cases, a transcranial approach is usually more advantageous, in particular in regard to watertight dural closure (see Chap. 4).

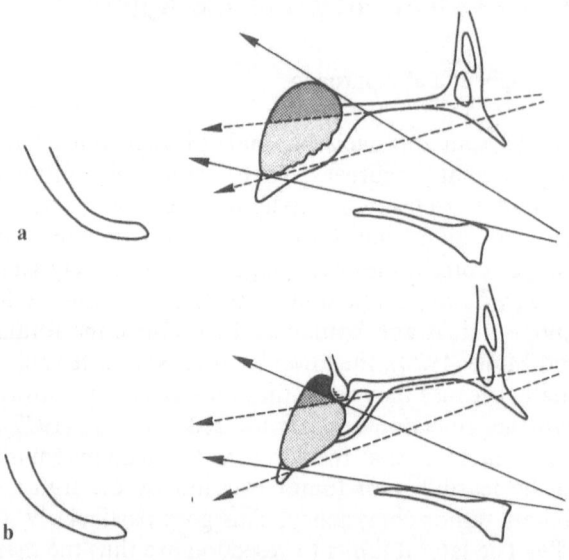

Summary: Transseptal-Transsphenoidal Approach

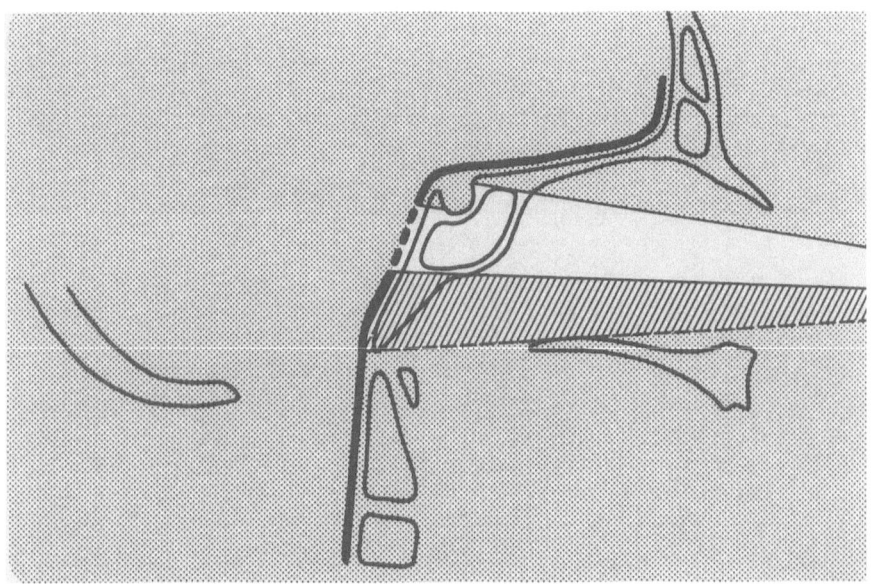

Indications: Lesions in the upper third of the clivus

Pituitary Adenomas – Chordomas

Advantages: Standard approach
                 Well tolerated

Disadvantages: Must be close to midline
                 Dural closure difficult
                 Danger of CSF leakage

Limits: Increasing difficulty in working on lower half of the clivus
                 Close to midline
                 Polycyclically suprasellar extension
                 Cavernous sinus (?)

Fig. 3.15a, b. Differences in the directions of approach in the transethmoidal (*dotted lines*) and transseptal (*unbroken lines*) variations of approach to lesions of the clivus **a** originating from the pituitary gland and **b** not originating from the pituitary gland. Tumor portions inaccessible by the transethmoidal approach are *shaded darker;* the portion of the dorsum sellae inaccessible by any of the approaches if the pituitary gland is preserved is *shaded very dark.* See also Fig. 3.5 b

Summary: Transethmoidal Approach

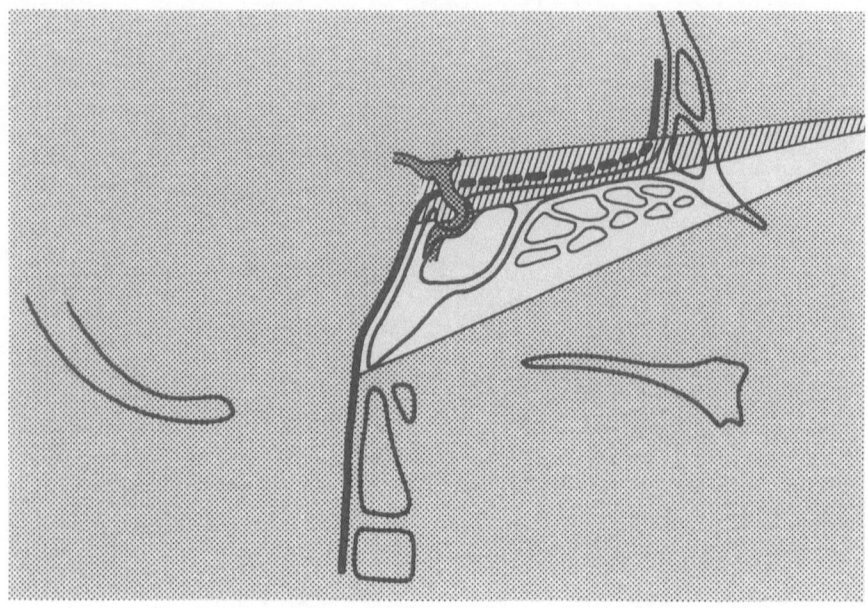

Indications: Extradural lesions of the ethmoid and sphenoid bones

Carcinomas – Pituitary Adenomas

Advantages: Extradural approach to the anterior skull base
Operation field not very deep

Disadvantages: Para-axial dissection (in unilateral surgery)
Cosmetic results unsatisfactory
Dural closure difficult

Limits: Frontobasal dura (?)
Supra- and retrosellar tumor parts inaccessible
Cavernous sinus (?)

Chapter 4: Transbasal Approach

Historical Survey

At first sight, it seems a long way from the anterior skull base to the clivus; however, the approach described below offers a wide range of variations and is fruitful ground for intraoperative cooperation with ENT surgeons. Nevertheless, this chapter is not meant to provide a detailed description of anterior skull base surgery, but rather to point out that it is possible to reach the clivus by an anterior skull base approach. The development of this approach was naturally founded on decades of experience in anterior skull base surgery (Unterberger 1958; Pertuiset 1955; Guiot and Derome 1966; Bonnal et al. 1961; Van Buren et al. 1968); it also comprises surgery of the orbit (Naffziger 1941; Brihaye et al. 1968; Schürmann and Voth 1972) and has been further enhanced by the progress made by Tessier in the field of craniofacial surgery (Tessier 1973; Tessier et al. 1967, 1973). In 1972, Derome and associates presented a comprehensive description of sphenoethmoidal surgery and laid the foundation of the "transbasal" approach to the clivus (Derome 1972; Derome and Guiot 1979; Arita et al. 1989).

Anatomy

Anatomy Relevant to Surgery

The extension of the *paranasal sinuses* and their projection onto the anterior skull base is very variable, so a diagram or table of measurements does not seem useful for present purposes. The medial orbital wall, which deviates very little from the sagittal plane and is situated 15 mm lateral to the median plane, is an important boundary (Fig. 4.1).

The most important dimensions pertaining to the anterior skull base are also shown in Fig. 4.1, except for the 18.5-mm difference in level between the endofrontal eminence and the lowest point of the olfactory groove.

The blood supply of the anterior skull base dura is mainly provided by the anterior and posterior *ethmoidal arteries* (see Fig. 4.1). According to Lang and Schäfer (1976), a third ethmoidal artery is found in 40% of cases. The posterior ethmoidal arteries exit from the orbits toward the ethmoid bone 5 mm (range 2–11 mm) short of the optic canal. The lateral portion of the anterior skull base dura projecting toward the lesser wing of the sphenoid bone is supplied

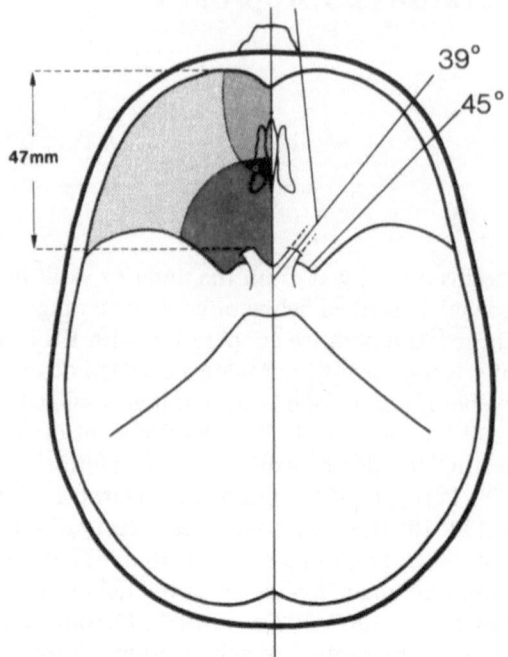

Fig. 4.1. The average distance between the interior aspect of the cranial vault and the optic canal is 47 mm. The angle between the lateral wall of the orbit and the median plane is 45°; the optic canal is situated at an angle of 39° from the latter. The medial wall of the orbit and the median plane form an angle of 6° (source: Lang 1981). The lefthand part of the diagram shows the arterial supply to the anterior skull base dura. The *light shading* marks the area supplied by the middle meningeal artery, and the *dark shading* the area supplied by ethmoidal artery. The branch of the internal carotid artery passing forward to the anterior clinoid process is not shown

by the middle meningeal artery. A branch coming directly from the internal carotid artery passes forward to the anterior clinoid process (Yaşargil 1984).

The angle formed by the two *optic nerves* and their intradural length largely depend on the form of the optic chiasm (prefixed, normal, or postfixed chiasm); the chiasmatic angle may vary from 50° to 80° (Renn and Rhoton 1975). The angle formed by the *optic canal* and the median plane is much more constant; according to Lang (1981) it is 30°. The osseous roof of the optic canal is markedly longer than its floor. A duplication of the dura, the *falciform ligament,* additionally covers the optic nerve from cranially; it is 3 mm (0.5 – 8 mm) long (Lang 1981). The optic canal is narrowest in the middle, where it is 4.6 mm wide (range 4 – 5 mm) (Lang 1981). A normal chiasm form is found in approximately 75% of cases. Prefixed chiasms occur in 10% of cases according to Renn and Rhoton (1975) or 9% according to Bergland et al. (1968). A more pointed chiasm angle, called postfixed chiasm, is found in 15% of all cases according to Renn and Rhoton (1975) or 11% according to Bergland et al. (1968). After removal of the anterior clinoid process and opening of the optic canal, the internal carotid artery is exposed from its lateral aspect. At the point where the *internal carotid artery* passes through the dura a fibrous ring around the artery attaches it to the osseous base. Only by severing of this fibrous ring can one reach loose connective tissue that allows blunt dissection without compromising the cavernous sinus (Perneczky et al. 1986; Perneczky et al. 1987, 1988; Knosp et al. 1988). In this method of dissection the caudal boundaries are formed by the cranial nerves passing through the superior orbital fissure (Knosp et al. 1988).

The 14-mm distance between the two optic canals (Lang 1981) has to be taken into account in the approach to the sphenoidal sinus and the clivus between the two optic nerves (see Fig. 4.4). Another important distance is that between the two carotid arteries, which is 7–16 mm in the area of the anterior knees and 8–22 mm at their horizontal segments. For more details regarding the lateral wall of the sphenoidal sinus, see Chap. 3 and Fig. 3.2.

In 41% of all cases the *ophthalmic artery* originates from the medial aspect of the internal carotid artery, in 32% from its anterior aspect, and in 25% from its lateral aspect, immediately after this artery passes inside the dura. In its further course the ophthalmic artery crosses below the canal segment of the optic nerve and crosses back over the orbital segment of the optic nerve from lateral to medial; this is found in 85% of all cases examined. In rare instances the ophthalmic artery originates extradurally (8% according to Renn and Rhoton 1975; 18% according to Engel 1975). According to Adachi the ophthalmic artery branches off the middle meningeal artery in 1% of cases, whereas the exact opposite, the middle meningeal artery branching off the ophthalmic artery, has been observed in 3% of cases (Adachi 1928, cited in v. Lanz and Wachsmuth 1979, p. 111). Anastomoses between the middle meningeal artery and the lacrimal branch of the ophthalmic artery are commonly observed as vestiges of embryonic development (Lasjaunias 1981; Knosp et al. 1987).

Key Steps of the Approach

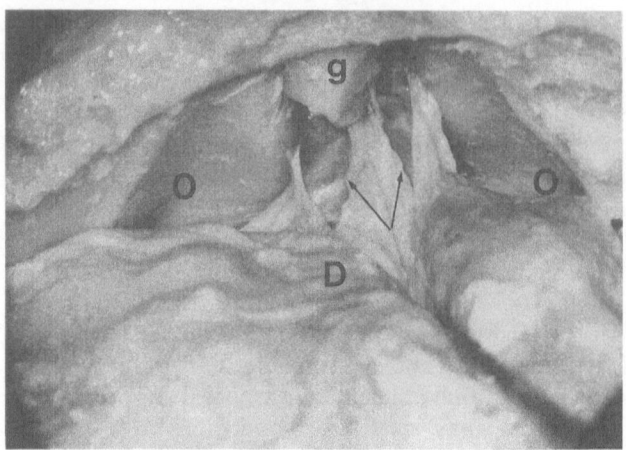

Fig. 4.2. After bifrontal craniotomy the anterior skull base is exposed extradurally. Dural lacerations in the area of the olfactory grooves are unavoidable (*arrows*). *D*, Anterior skull base dura; *O*, orbital roofs; *g*, crista galli

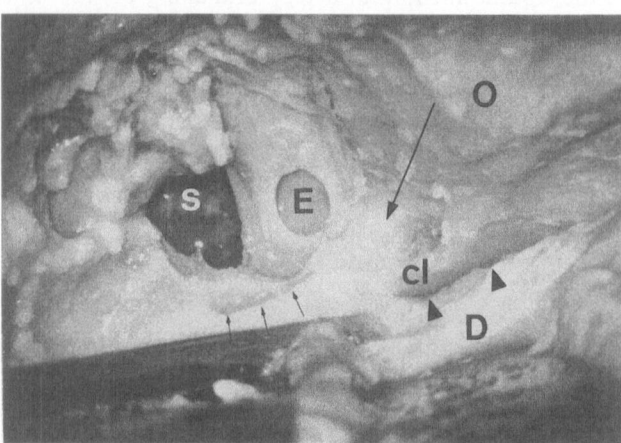

Fig. 4.3. Start of median dissection toward the clivus with exposure of the right orbital apex and after opening of the sphenoidal sinus (*S*). A large posterior ethmoid cell (*E*) borders the optic canal on the right. Part of the right orbital roof (*O*) and the anterior clinoid process (*cl*) have been removed. *Small arrows*, Jugum sphenoidale; *long arrow*, right optic nerve; *D*, anterior skull base dura; *arrowheads*, edge of the right sphenoid wing, see also Fig. 4-5

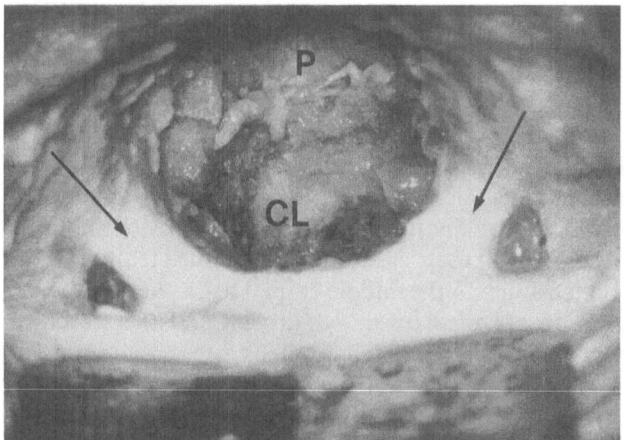

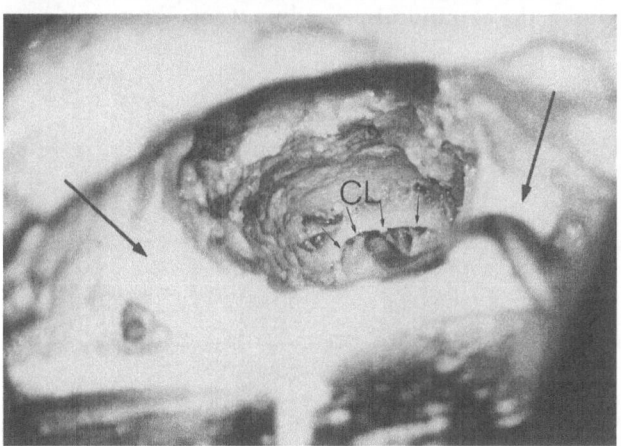

Fig. 4.4. a View of the osseous clivus (*CL*) and the epipharynx (*P*) between the two optic nerves (*arrows*). The ethmoid and nasal septum have been resected, as have the medial orbital walls and parts of the orbital roof. Note the pneumatization of the left anterior clinoid process. **b** View of the lowest part of the clivus (*CL*). The basal clival dura is detached from the osseous rim with a dissector, thus exposing the ventral margin of the foramen magnum (*small arrows*). *Large arrows,* optic nerves

Surgery

Technique

A coronal skin incision should be placed well beyond the hairline with a view to obtaining the largest possible galeaperiosteal flap. Bifrontal craniotomy, which usually includes incision of the frontal sinuses, must reach the base. Even greater flexibility is attained if the craniotomy is continued temporally on one side, so that tumor extension into the middle cranial fossa can be exposed. The wide range of variation afforded by this approach and the structures accessible to it are illustrated in Fig. 4.5. The next steps after craniotomy has been completed are determined by the relation between the lesion and the dura, i. e., whether one will work primarily extradurally or whether one wants to continue extradural surgery only after removal of intradural parts of the tumor and dural closure. We favor the extradural procedure whenever possible.

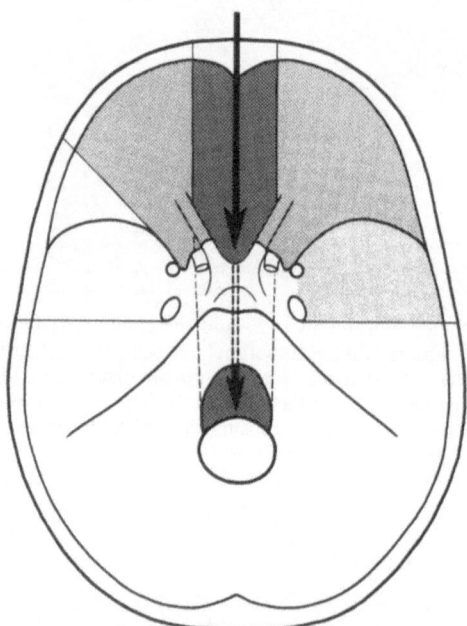

Fig. 4.5. Accessibility of the anterior skull base after bifrontal craniotomy (*medium shading*). Concomitant pterional craniotomy renders the middle cranial fossa up to the internal carotid artery and the foramina rotundum and ovale accessible (*light shading*). After resection of the ethmoid, a transbasal approach between the optic nerves to the sphenoidal sinus, the clivus, and the topmost cervical vertebrae (*dark shading*) is possible. Thus, important structures such as the internal carotid artery and the optic nerve can be exposed from ventrally, medially and laterally

The dura must be detached from the anterior skull base after that, the almost unavoidable dural lacerations in the area of the cribriform plate of the ethmoid have to be patched. Via this bifrontal approach it is possible to reach the orbital roofs and optic canals on both sides, the entire ethmoid, the medial orbital walls, and the sphenoidal sinus. As soon as the optic nerves have been detached from the osseous structures, the anterior aspect of the sella turcica and then the clivus become accessible between them. In the extreme depth, the anterior arch of the atlas and vertebral body C2 can be reached. Only the portion of the clivus shielded by the sella turcica remains inaccessible. The visual angle can be slightly improved if the sellar floor is drilled off. If the tumor has not displaced the anterior parts of the carotid siphon, these may restrict access to the clivus. Renn and Rhoton (1975) reported that C3 segments of the internal carotid arteries (anterior siphon knee) may come as close to each other as 4 mm. Such problems should, however, have been clarified prior to surgery by angiography or MRI. If the lesion requires additional pterional craniotomy, the procedure is as follows:

The angle of approach is changed by 30°. In this way, extradural dissection along the lesser wing of the sphenoid bone and its complete resection are possible. Thus, the lateral side of the internal carotid artery is reached at the area of entry into the dura after resection of the anterior clinoid process. The anterior aspect of the C3 segment is not usually surrounded by the veins of the cavernous sinus; a risk of opening the cavernous sinus is only imminent in the portion dorsal to the anterior siphon knee, behind the anterior clinoid process (Perneczky et al. 1985).

The lesser wing of the sphenoid bone having been removed, the orbital fissure is opened. Further basal resection is obstructed by the oculomotor, trochlear, abducent and the ophthalmic nerve passing through the superior orbital fissure, if they are still intact. If craniotomy extends far enough laterally, the anterior portions of the middle cranial fossa can be exposed and the osseous structure can, if necessary, be resected as far as the foramina rotundum and ovale (Fig. 4.5). At this point, the limits of this approach have been reached. In extreme cases, the orbits, optic nerves, both internal carotid arteries, the pituitary gland, the clivus and – in the very depths – the topmost cervical vertebrae can be exposed. If the tumor invades the dura or there are large intradural portions of tumor, we start by resection of the latter. When this has been completed, the dura must be sealed off watertight. Only then should the extradural tumor portions be attended to, or they should even be removed in a second-stage operation if prolonged surgery time, extensive loss of blood and other factors make this seem advisable. If a tumor involves the dura, it is often necessary to cover large dural defects. As the petiolated galaperiosteal flap has been prepared for covering the base adjacent to the paranasal sinuses, either fascia lata or a free galeaperiosteal flap taken from the posterior parietal scalp should be used for the dural defect itself; lyophilized dura may also be used. We consider it more important to use the well-vascularized, vital layer of the petiolated galeaperiosteal flap to seal off the paranasal sinuses. After closure of the dura, the procedure is the same as for primarily extradural excision.

Reconstruction of the bone by means of an autologous bone graft or methyl-acrylate can only be carried out if the paranasal sinuses have been sealed off completely. Vital autologous tissues, such as petiolated galeaperiosteal flaps, petiolated temporal muscle, or fasciae of the temporal muscle as well as intact nasal mucosa are used for cover. Insufficient sealing off inevitably leads to infection and prevents the bone graft from settling. It is essential that the nourishing layers are placed adjacent to the grafts for sufficient and early bone nourishment and to prevent hematoma in cavities. Performing the reconstruction in several, preferably independent compartments may be advantageous, as will be shown in the following case report. With regard to the bone material to be used cleft ribs lend themselves well to the reconstruction of the supraorbital arch. Ribs may also be used to reconstruct the osseous clivus, but the iliac crest is apparently better suited for this purpose due to its sturdiness. If only the ethmoid is to be covered, fascia lata alone or a portion of the internal tabula of the calvaria is sufficient.

Case Report

This 23-year-old woman was afflicted with fibrous dysplasia, for which she had twice undergone surgery in her home country. She suffered from increasing headache, diplopia caused by increasing protrusion of the eyeballs, and discrete hypesthesia V_1 on the left side. Visual acuity and fields of vision were normal. Conventional tomography gave impressive enough results in revealing the extension of the fibrous dysplasia from the left frontal sinus to the clivus. Nevertheless CT was used to complement the survey of severe destruction of the osseous base. The ethmoid as well as the entire sphenoid bone and the clivus were affected. The skin incision, based on previous surgery, was extended in the form of a coronal skin incision, followed by bifrontal craniotomy right to the bottom of the frontobasis. In addition, left temporal craniotomy was performed. Subsequently, the lesser wing of the sphenoid bone was drilled off as far as the anterior clinoid process. Thus, the superior orbital fissure was opened cranially and laterally. As soon as sound bone was reached in the area of the middle cranial fossa, lateral dissection was terminated.

In the second part of the operation the heavily thickened anterior skull-base was drilled off from the supraorbital rim to the tuberculum sellae. The affected osseous structures surrounding the optic nerves were removed. Now the sphenoid body situated at the midline was drilled out between the two optic nerves. In the further course of dissection the pituitary gland was reached. After removal of the anterior wall and the floor of the sella, the clivus was attained and the affected bone there removed as well. The entire ethmoid bone, large portions of the sphenoid bone, and the medial and superior walls of the orbit were thus resected in the ENT part of the operation. In treating such extensive

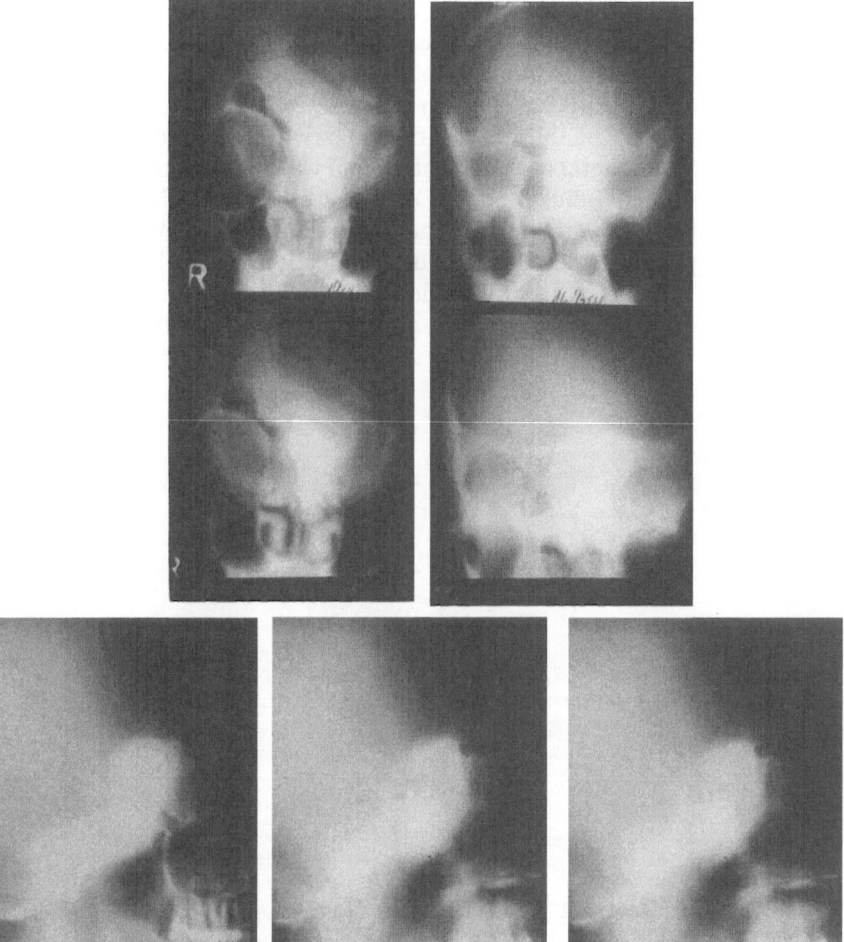

Fig. 4.6. Conventional tomography: frontal and lateral views. The severe osseous destruction starts at the left frontal sinus and continues as far as the clivus. The osseous defect on the frontal left stems from previous surgery. The lateral view shows that the sphenoidal sinus is no longer pneumatized, whereas the sella turcica and the dorsum sellae are hardly affected

pathological processes, the reconstruction of the osseous base is definitely a very important stage. When methylacrylate is used for reconstruction of the base, it is crucial to cover the graft adjoining the nasal cavity with a well-vascularized layer in order to prevent infection.

Although the risk of infection is admittedly high, factors favoring this way of reconstruction are moldability and avoidance of an additional operation at the donor site. Methylacrylate was used in this case because a sufficient quantity of vital nasal mucosa was available to cover the reconstruction of the medial orbital wall. A second protective layer placed on the medial orbital wall and the orbital roof consisted of a petiolated portion of temporal muscle and its fasciae. The frontobasis was sealed off by means of a galeaperiostal flap. The latter was covered with a thin plate of methylacrylate serving to reconstruct the anterior skull base, which was then covered with dura from the anterior skull base. The third piece of methylacrylate was independently used to reconstruct the frontal calvaria. The wound was closed in the usual manner.

Infection occurred due to wound dehiscence of the original surgery scar. However, only the calvaria portion had to be removed. The other two portions of methylacrylate healed well in spite of the infection, as they had been covered with at least one layer of vital tissue on all sides. The calvaria defect was covered some months later without complications and with good cosmetic results.

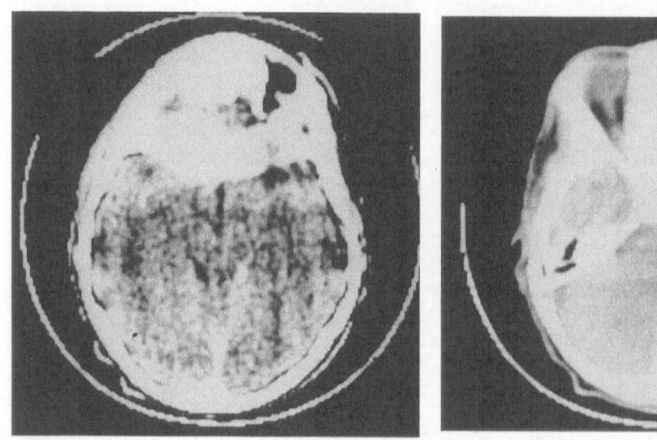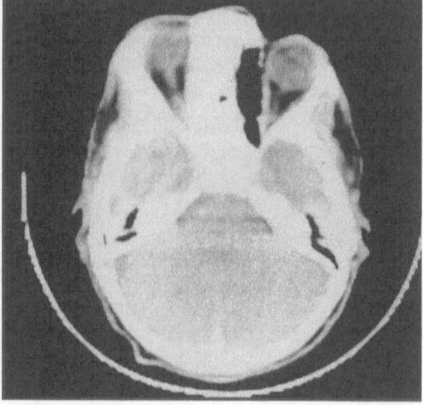

a   b

Fig. 4.7 a, b. Preoperative CT shows the severe osseous deformation of the left medial and left lateral orbital walls. The left eyeball also protrudes slightly

Fig. 4.8 a−c. Intraoperative site. **a** Destruction of the frontal vault of the cranium after dissection of the skin flap; the two *arrows* mark the thickness of the cranial vault. **b** Situation after resection of the deformed bone: both optic nerves (*arrows*) have been detached from the bone. *O*, orbital contents after resection of medial and lateral orbital walls and orbital roof. *S*, Remnants of the wing of the sphenoid bone. View of the deep resection cavity (*X*) between the optic nerves, reaching to the clivus. **c** A petiolated pericranial flap (*F*) is prepared to cover the large osseous frontobasal defect

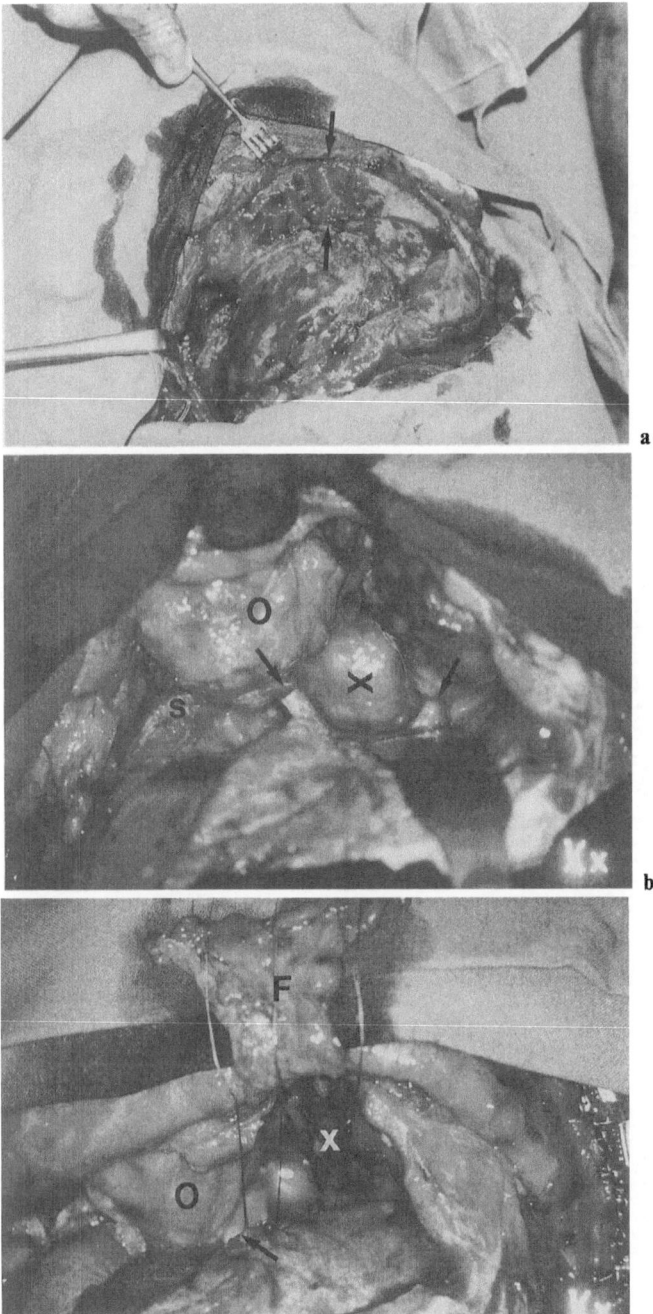

a

b

c

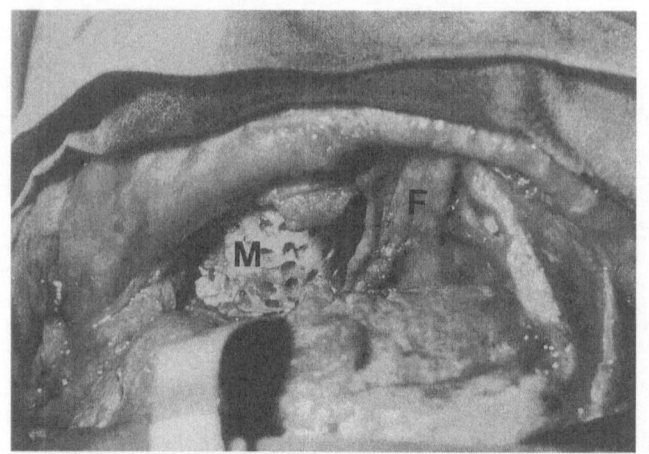

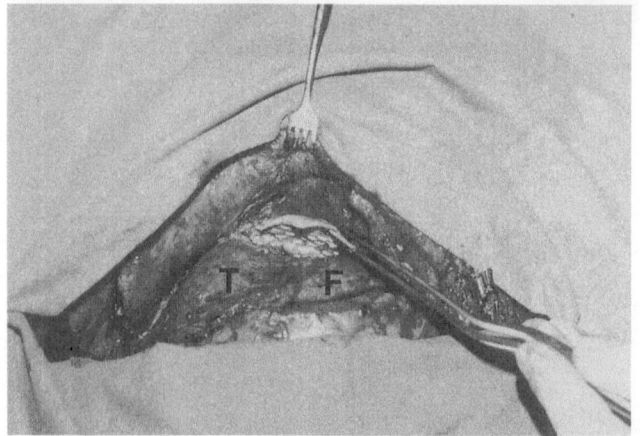

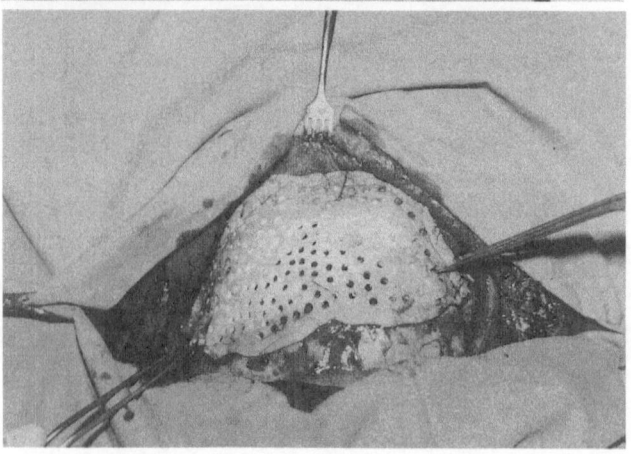

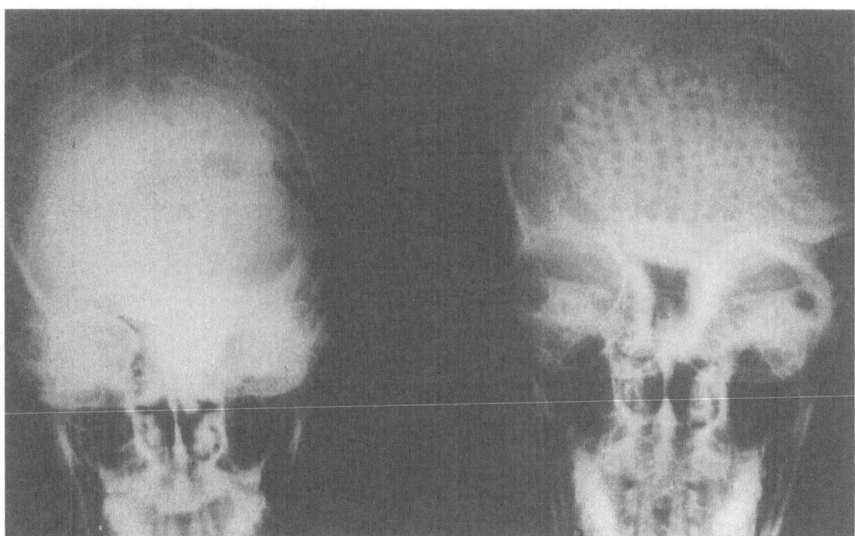

Fig. 4.10. Preoperative and postoperative cranial X-rays compared

Frequently Encountered Lesions

The transbasal approach is indicated for lesions which start from the sphenoid bone itself and encroach on the ethmoid bone, the anterior skull base, and the clivus. Such processes may be *meningiomas* growing invasively through the skull base. They are in addition space-occupying lesions, intracranially in most cases, either in the anterior cranial fossa, such as olfactory groove meningiomas, or meningiomas of the tuberculum sellae. They may also develop toward the lesser wing of the sphenoid bone in the form of meningiomas of the sphenoid ridge. In the latter case, they grow into the lesser wing of the sphenoid bone, infiltrate the cavernous sinus, and thus also encroach on the body of the sphenoid bone. They may, however, also encroach on the dorsum sellae along the lateral wall of the cavernous sinus and reach the clivus from there. Hyperostotic bone adjacent to meningiomas represents infiltrative tumor growth (Bonnal et al. 1980). In keeping with the principle of radicality, the hyperostotically deformed bone should be removed.

Fig. 4.9. a Reconstruction of the frontobasis by methylacrylate. Reconstruction of the medial orbital walls and the orbital roof on the left (*M*) and covering of the graft with a galcaperiosteal flap (*F*), which is sutured with an additional petiolated flap of temporal muscle (*T*). **b** Fitting in a plate of methylacrylate to reconstruct the osseous frontobasis. **c** Final reconstruction of the frontal vault of the cranium

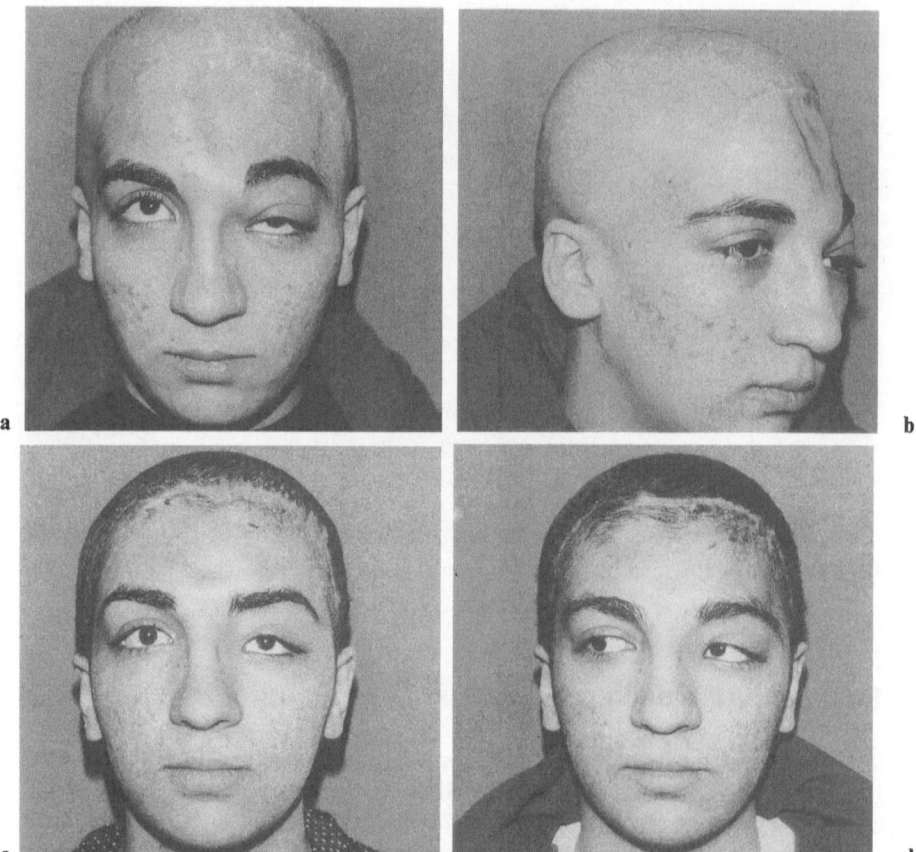

Fig. 4.11 a–d. Patient before surgery (**a, b**) and cosmetically satisfactory postoperative result (**c, d**)

Chordoma, chondroma and chondrosarcoma develop in the sellar and parasellar regions. Their growth destroys the bone without at first invading the dura and are they often only discovered when they have reached a considerable size, by which stage the entire body of the sphenoid bone and parts of the ethmoid bone may be affected. Although radiotherapy is carried out in many cases (Poppen and King 1952; Zoltan and Fenyes 1960; Krayenbühl and Yaşargil 1975; Raffel et al. 1985), the only real chance for cure is surgery as radical possible (Schisano and Tovi 1962; Falconer et al. 1968). Due to the lack of symptoms at the early stage, patients often undergo surgery at a stage in which the tumor has become so large that surgery of enormous extent, such as transbasal surgery, is necessitated for the sake of radical excision – if this is possible at all.

Due to its growth *fibrous dysplasia* often destroys the entire skull base and is therefore a further indication for the transbasal approach (Derome 1972; Draf and Samii 1977).

Advantages and Disadvantages of the Approach

The transbasal approach is the only one that makes it possible to reach lesions extending from the anterior skull base to vertebral body C2 within one operation. Moreover, it offers variability in the direction of approach; tumors engulfing important structures may thus be removed from two sides. Unlike in the other "anterior" approaches to the clivus, dissection is not limited laterally. Moreover, there is enough space to resect even compact lesions – if need be, by means of a drill. Dural damage or resection does not represent an additional risk because better view allows accurate closure of the leak or dural replacement is feasible in a controlled fashion; the intradural portion can thus also be reached within one operation. The approach is particularly suited for portions of tumor encroaching on the ethmoid including the orbital walls and/or the nasal cavity, as they can be resected along with the rest in one operation and do not require additional transethmoidal surgery.

The reconstruction of the frontobasis by autologous bone graft is rather intricate, but it is a decisive stage in the operation. First, the dura adjoining the paranasal sinuses must be closed watertight by means of an appropriate galeaperiosteal flap. The dura may also be covered by fascia lata, a free periosteal flap from the parietal bone, or an epidermal flap (Derome 1972). Lateral portions of the intact anterior skull base dura may also be detached and sutured above the defect like the wings of a double door (Draf and Samii 1986). In a similar manner, the superficial fascia layer of the temporal muscle may be mobilized in a petiolated flap and turned over the anterior skull base. In large osseous defects of the anterior skull base such maneuvers alone are insufficient to prevent postoperative encephalocele and protrusion of the bulb. Anterior skull base reconstruction thus requires reconstruction not only of the dura, but also of the bone and the mucosa (Derome 1972). When methylacrylate is used, which is easier to mold, absolute sealing off of the nasal cavity is essential for healing without infection. The graft should be covered by a petiolated flap (Draf and Samii 1986). The advantage of moldability has to be weighed against the disadvantage of the high risk of infections. An alternative is the use of autologous pelvic bone from the posterior iliac crest (Derome and Guiot 1979). A portion of the internal tabula may be taken from the calvaria to cover minor defects. The use of cleft ribs has also been recommended (Derome 1972). In the latter case, the cortical layer of the bone should adjoin the paranasal sinuses, and parts of the nasal mucosa or a vascularized flap should adjoin the bone to guarantee the best healing possible.

Summary: Transbasal Approach

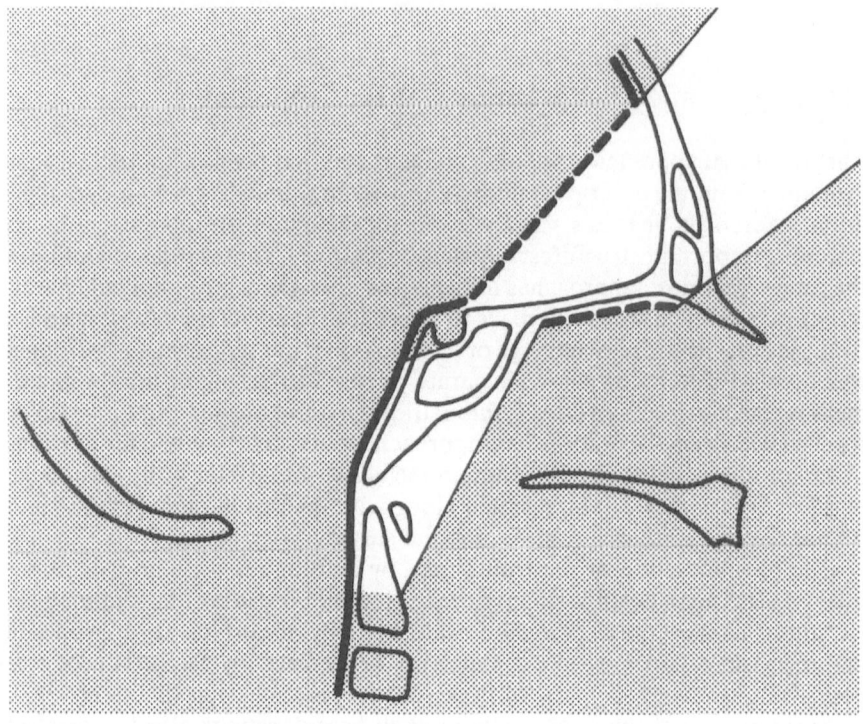

Indications: Lesions of the sphenoethmoid bone of extra- and intradural extensions

Fibrous Dysplasia – Meningiomas – Chordomas

Advantages: Extra- and intradural tumor portions
Safe dural closure

Disadvantages: Large-scale surgery
Painstaking reconstruction

Limits: Topmost clival regions (dorsum sellae)
Caudal to C2
Cavernous sinus (?)

Chapter 5: Laterobasal Approach

Historical Survey

The development of ENT surgical techniques, which were first limited to the petrous bone, soon led to contact with neurosurgeons. For decades, the cerebellopontine angle provided fertile soil for discussion. During the past few years, further development of surgical techniques has shifted emphasis to the petrous apex and clivus, and as the surgical techniques approached one another, greater cooperation between neurosurgeons and ENT surgeons ensued. Early attempts at reaching the cerebellopontine angle by means of a combined suboccipital-transpetrosal approach were abandoned because of fatal hemorrhaging from the sigmoid sinus (Fränkel 1904; Borchardt 1905; Marx 1913; Bailey 1939). In 1966, Hitselberger and House returned to this option and presented this approach. They performed it almost exclusively in two stages and resected the tumor during the second, translabyrinthine stage.

The labyrinthine segment of the facial nerve represented an obstacle in the access to petrous apex and clivus in translabyrinthine operations. It thus also consituted the anterior limit to the translabyrinthine approach. More extensive access to the clivus was only rendered possible by mobilization of the nerve from its osseous facial canal and its displacement. This type of extension was termed the "transcochlear approach" to the skull base by House and Hitselberger in 1976. Now the internal carotid artery represents the anterior limit to resection. Fisch (1978) together with Pillsbury (Fisch and Pillsbury 1979), went one step further: having displaced the facial nerve, they dissected along the internal carotid artery as far as where it enters the cavernous sinus; thus, they were able to expose the entire petrous segment of the internal carotid artery (1982). Fisch first and foremost reached the jugular foramen by an approach termed the "infratemporal fossa approach," but he also touched parasellar and clival regions after mandibular condyle resection (Kumar and Fisch 1983). He limited himself to extradural tumors, however: i.e., intradural portions of tumor had to be removed in a second, purely neurosurgical stage. "Radical operations" on malignant growths in the temporal bone led to even more extensive surgery; the transotic approach of Jenkins and Fisch (1981), which included more or less complete resection of the temporal bone with sacrifice or preservation of the internal carotid artery, represents a further extension of the approach described above (Parson and Lewis 1954; Crabtree et al. 1976; Gacek and Goodman 1977; Glasscock 1974, 1978, 1983, 1985; Lewis 1983). In this was, exposure of the parasellar, para- und retropharyngeal spaces, the clivus, the foramen magnum, the atlas, and the axis became possible.

Anatomy

Anatomy Relevant to Surgery

The *facial nerve* ist of focal importance for approaches through the petrous bone. From its intracisternal segment (cf. also the lateral suboccipital approach Chap. 7), the facial nerve enters the Fallopian canal above the transverse crest at the fundus of the internal auditory meatus. The subsequent labyrinthine segment passes forward as far as the geniculate ganglion; it is the narrowest part of the facial nerve canal. In this segment, the facial nerve is only covered by a very delicate layer of periosteum. Due to this, the bottleneck in the osseous structures, and a critical vascularization pattern (May 1986), this is the part of the facial nerve at greatest risk in its intratemporal course. After the greater petrosal nerve has branched off the facial nerve at the geniculate ganglion, taking parasympathetic fibres of the sensory root of the facial nerve to the lacrimal gland, the facial nerve turns dorsally and enters its tympanic segment, which is where the perineural sheath starts to develop. The facial nerve passes in a dorsal direction between the lateral semicircular canal and the stapes (see Fig. 5.2), then takes a less marked turn in a caudal direction and reaches the stylomastoid foramen along the anterior aspect of the mastoid bone. Variations in the course of the facial nerve in the tympanum and mastoid bone have been summarized by Helms (1981). It is still in this segment – called the mastoid segment – that the chorda tympani branches off the facial nerve, passes through the tympanic cavity, and subsequently terminates at the tongue.

The facial nerve is supplied by branches of the labyrinthine artery in the internal auditory canal and the adjoining labyrinthine segment; its tympanic segment is vascularized by the petrosal branch of the middle meningeal artery, which reaches the geniculate ganglion along the greater superficial petrosal nerve. The stylomastoid artery supplies the mastoidal segment of the facial nerve. In the labyrinthine part, there is one critical spot in the vascularization of the nerve due to the lack of anastomosing arcades (May 1986).

The *labyrinth* is situated between the facial nerve and the posterior aspect of the petrous bone. The lateral semicircular canal is located a little above the tympanic segment of the facial nerve; it is reached first after mastoidectomy. This is also the safest position at which to identify the facial nerve. The anterior semicircular canal, elevating the arcuate eminence at the surface of the petrous bone, is situated at right angles to the ridge of the petrous bone. The labyrinthine segment of the facial nerve passes only a few millimeters short and parallel to the anterior semicircular canal. The posterior semicircular canal is situated in the plane of the posterior surface of the pyramid, lateral to the internal auditory canal (Firbas 1985).

The *chochlea* is located at the apex of the petrous bone between the labyrinthine segment of the facial nerve and the internal carotid artery. No more than a few millimeters separate it from the two structures. The bone separating the cochlea from the internal carotid artery usually measures no more than 1 mm (House 1968; House and Hitselberger 1969, 1976). The zone in which the cochlear nerve can be seen entering at the fundus of the internal auditory canal outlines the basal turn of the cochlea. This zone, called the cochlear area, is separated from the Fallopian canal by the transverse crest. In approaches from the petrous bone, a vertical crest, also named "Bill's bar", becomes visible earlier. It separates the superior vestibular nerve from the facial nerve, which is situated ventrally.

Key Steps of the Approach

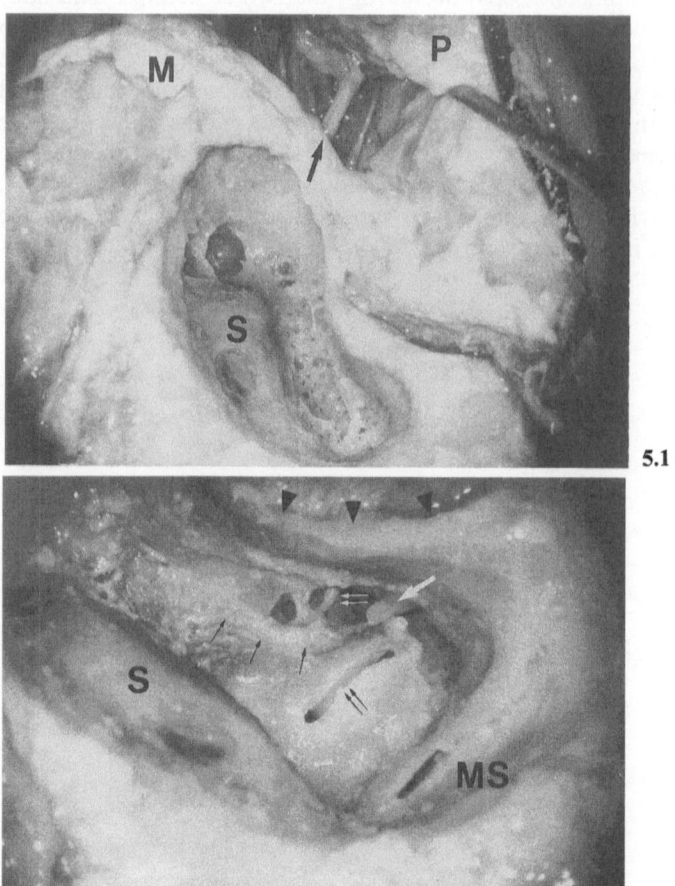

5.1

5.2

Fig. 5.1. After the facial nerve (*arrow*) has been exposed outside the stylomastoid foramen, mastoidectomy is begun (dissection is carried out on the left). The sigmoid sinus projects from dorsally and is still covered by a thin osseous lamella (*S*). *M*, Tip of the mastoid; *P*, parotid gland

Fig. 5.2. Start of labyrinthectomy after exposure of the floor of the middle cranial fossa. The lateral semicircular canal has already been opened (*double black arrows*). The facial nerve turns into its mastoid segment (*ms*) below the lateral semicircular canal (*single black arrows*). *White arrow*, Incus; *arrowheads*, Osseous boundary of the external auditory meatus; *double white arrows*, stapes. *S*, sigmoid sinus

5.3

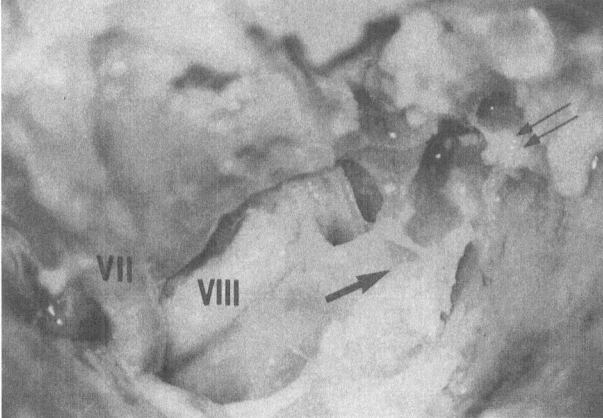

5.4

Fig. 5.3. After labyrinthectomy the internal auditory meatus (*black arrow*) is openend. Thus, the intrameatal and tympanic segments (*white arrows*) of the facial nerve are exposed. The labyrinthine segment is not visible in this projection. *White double arrows*, geniculate ganglion

Fig. 5.4. The fundus of the interal auditory canal with the facial nerve already transposed anteriorly (*VII*). The basal turn of the cochlea has been opened, showing the modiolus. Note the rotation of the cochlear nerve (*VIII*). *Arrow*, facial nerve canal; *double arrow*, stump of the transected greater petrosal nerve

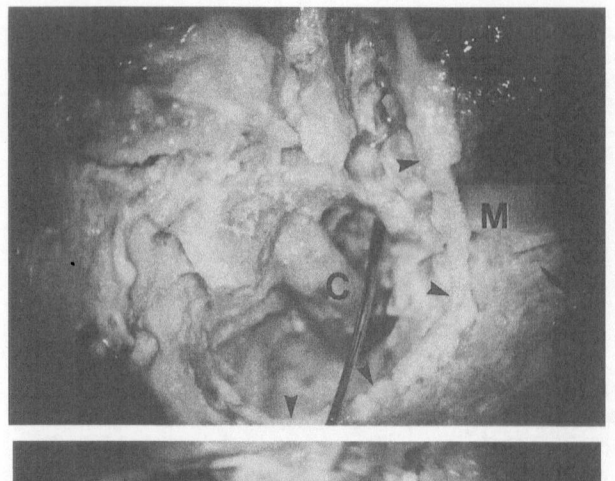

5.5

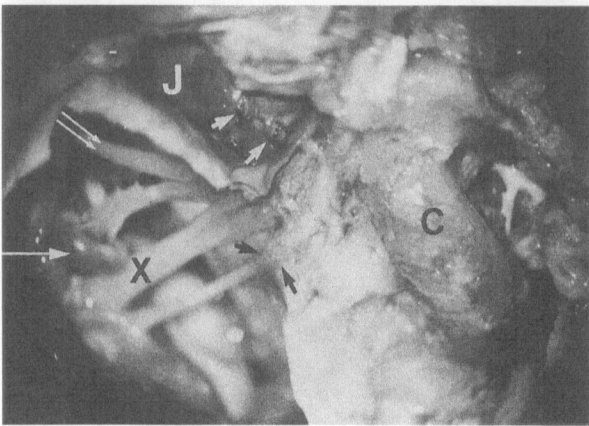

5.6

Fig. 5.5. The apex of the petrous bone has been removed, thus opening the intrapetrous carotid artery (*C*), and the Eustachian tube (probe). *Arrowheads,* transposed facial nerve; *M,* mandibular joint

Fig. 5.6. By further bone resection the jugular bulb (*J*) is reached and opened. This brings the caudal cranial nerves into view: the glossopharyngeal, the vagus (*X*) and the accessory nerve (*long white arrow*). *White double arrow,* loop of PICA between fibers of the vagus nerve. Inside the jugular bulb, the orifices of the inferior petrosal sinus can be seen (*white arrows*). Note the separate dural porus (*black arrows*) for the glossopharyngeal nerve

Surgery

Technique

Translabyrinthine Approach

The skin is incised in a semicircle behind the ear. Then, the mastoid is opened and the sigmoid sinus and the dura at the posterior aspect of the apex of the petrous bone and also the temporobasal dura are exposed (Fig. 5.2). Hemorrhage from the petrosal sinus, which may occur at this stage, is often unpleasant and has to be packed. After the tympanic segment of the facial nerve has been identified and exposed below the lateral semicircular canal, the labyrinth is drilled off. In this way the internal auditory canal is reached and exposed at its upper circumference as fas as the porus. When the facial nerve has been identified at the fundus of the internal auditory meatus, exposure of the labyrinthine portion of the nerve is possible. The anterior limit of the translabyrinthine approach has been reached when the entire temporal segment of the facial nerve has been exposed. When the dura is opened medial to the sigmoid sinus, this approach is suited for reaching small lesions of the cerebellopontine angle, but not for lesions of the apex of the petrous bone and the clivus.

Transcochlear Approach

The decisive step to extend the approach lies in the skeletization of the facial nerve from the cerebellopontine angle to the stylomastoid foramen. Both the greater petrosal nerve and the chorda tympani have to be severed to allow detachment of the nerve from the Fallopian canal. Only then may the facial nerve be displaced dorsally, thus allowing free access to the cochlea. After removal of the ossicles, the cochlea is drilled out of the petrous bone. No more than an thin osseous lamella separates the basal turn of the cochlea from the petrosal segment of the internal carotid artery, which marks the anterior limit to the transcochlear approach. Resection of the bone as far as to the inferior petrosal sinus and the jugular bulb results in funnel-shaped access to the petrous apex and the clivus. The funnel ends a little below Meckel's cavity. The limits to the transcochlear approach are marked as follows: ventrally by the internal carotid artery, cranially by the superior petrosal sinus, caudally by the jugular bulb. Lesions also involving the infratemporal fossa and/or the paraphyaryngeal space require radical extension of the approach.

Infratemporal Fossa Approach

The skin incision is performed parallel to the sternocleidomastoid muscle, with a Y-shaped division a little below the ear. The ear ist cut at the external auditory meatus, the auditory meatus is closed off and turned up with the skin flap (Fig. 5.10). An alternative skin incision leads along the anterior rim of the sternocleidomastoid muscle to the angle of the mandible, then dorsally as far as behind the mastoid and parallel to the auricle, describing an arch in a temporal direction; in this case, the ear cut at the external auditory meatus has to be turned forwards. The internal carotid artery and jugular vein are dissected, the external carotid artery is ligated at its origin, and the caudal cranial nerves are identified, first in the cervical region, and then as far as the skull base, if possible, but in any event as far as the tumor boundary. The hypoglossal nerve crosses the operation site and requires special attention.

The first stage of dissection is completed by exposure of the inferior aspect of the tumor and of the caudal cranial nerves in the cervical region. Subsequently, the facial nerve, which is of crucial importance in this approach, is exposed. After it has been exposed at the stylomastoid foramen, dissection of the nerve along its entire temporal course is carried out, allowing anterior displacement. Depending on the size of the tumor, partial parotidectomy and further exposure of the peripheral branches of the facial nerves may be required.

The next step, craniotomy of the posterior and middle cranial fossa, is performed with a view to exposing the transverse and sigmoid sinuses and the intracranial boundary of the tumor. If necessary, the transverse and/or sigmoid sinuses are ligated, following which tumor extirpation can be embarked on. The internal jugular vein is ligated and transected at the neck. Dissection proceeds along the internal carotid artery; step by step, all the branches of the external carotid artery supplying the tumor are ligated and severed. The cranial nerves at the jugular foramen are exposed under the microscope. They may be dissected from intradurally as well as from the neck. The bone over the jugular bulb and the intrapetrosal internal carotid artery are drilled out and the tumor is gradually removed. The affected dura has to be resected along with the tumor. In large tumors reaching the parasellar region, the mandibular condyle has to be resected so that the tumor can be traced as far as behind the posterior part of the cavernous sinus. Major hemorrhage from the inferior or superior petrosal sinus must be controlled by packing and compression. The dural defect is sealed off watertight with lyophilized dura. The large osseous defect is covered with parts of the sternocleidomastoid muscle and portions of the exposed temporal muscle (Denecke 1966). These muscle portions may also be used to seal off the Eustachian tube in order to prevent CSF leakage and wound infection.

Case Report

A 23-year-old man consulted an ENT specialist because of a secretion from his right ear. Examination revealed a bluish tumor on the right tympanic membrane. Progressive loss of hearing had developed and a pulsating mass bulged into the oropharynx (Fig. 5.7). As for the cranial nerves, there was discreet paresis of the hypoglossal nerve on the right and transitory diplopia when the eyes were turned to the far right. Otherwise, there were no neurological deficits. Plain skull films showed a shadow of soft tissue density in the tympanic cavity reaching to the external auditory meatus, accompanied by destruction of the right apex of the pyramid. Angiography and CT revealed a glomus jugulare tumor of quite extraordinary size (Figs. 5.8, 5.9). Intraoperative cooperation between a neurosurgical and an ENT team working alternately made it possible to resect the tumor in one operation.

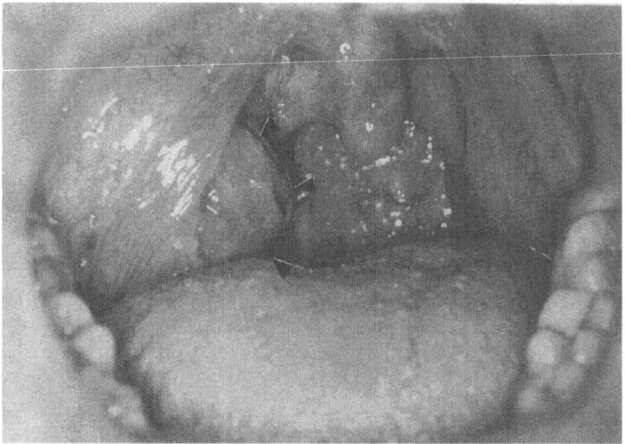

Fig. 5.7. The bulge of the tumor in the right oropharynx (*arrows*)

The ENT team performed a Y-shaped skin incision and turned up the ear so as to expose the blood vessels of the neck as well as the facial, glossopharyngeal, vagus, accessory, and hypoglossal nerves (Fig. 5.10). The hypoglossal nerve, which was exposed as far as the tongue in order to attain the best possible mobilization, required especially careful dissection. For sufficient mobilization, the facial nerve was followed far into the parotid gland, which was partly extirpated. In the ensuing petrosectomy, the facial nerve was exposed along its entire course and mobilized. The size of the tumor necessitated removal of the mandibular condyle and the adjacent portions of the zygoma. The following, neurosurgical stage of the operation started with trepanation of the middle and posterior cranial fossae with a view to exposing the sigmoid and transversal sinuses, which were both infiltrated by the tumor. After excision of the affected dura, the transverse sinus was ligated and severed. Tumor resection was begun only afterwards. The strategy followed was ligating the large venous blood conduits – the transversal and sigmoid sinuses – prior to the internal jugular vein. Tumor dissection was then embarked on from the neck and continued along the internal carotid artery to the skull base, with the cranial nerves being preserved. The vagus nerve had been split into individual

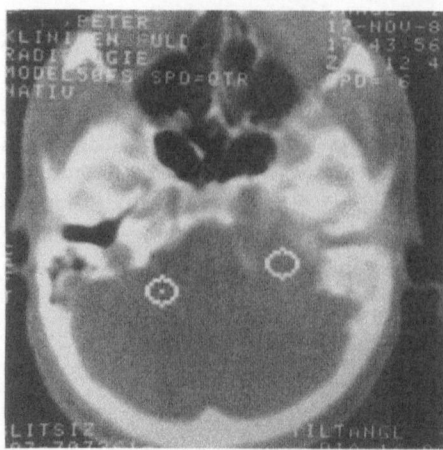

Fig. 5.8. CT clearly shows the destruction of the apex of the petrous bone, extending from the osseous external auditory meatus almost to the middle of the clivus

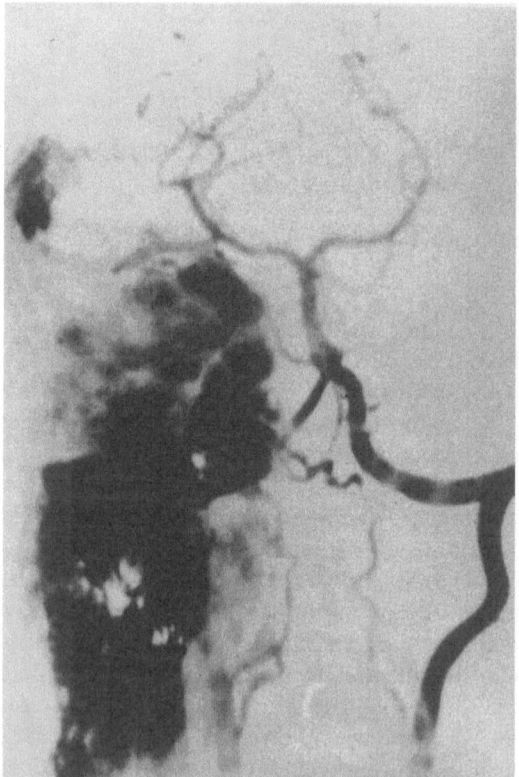

Fig. 5.9. Left-side vertebral angiography shows the typical vascularization of an extensive glomus jugulare tumor, which extends a considerable way intracranially and reaches the carotid bifurcation in the cervical region

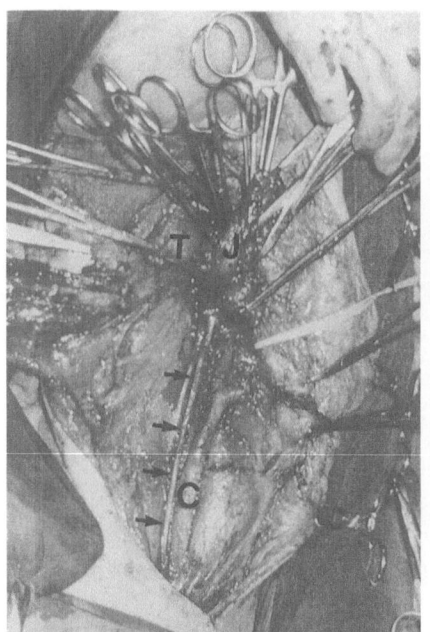

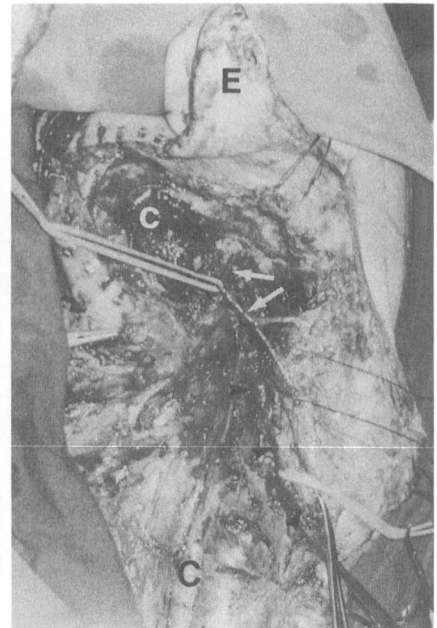

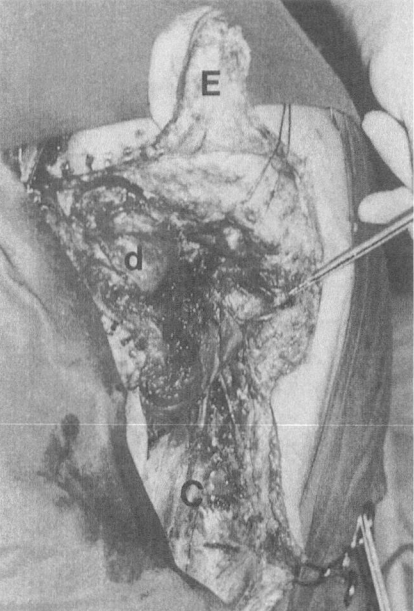

Fig. 5.10. a Intraoperative situation at the beginning of tumor dissection: exposure of the common carotid artery (*C*), vagus nerve (*arrows*), and snared hypoglossal nerve (*XII*). The external carotid artery has already been ligated and severed. The tumor (*T*) is dissected from caudally along the internal carotid artery, the jugular vein (*J*) having been severed. **b** After tumor extirpation. *Arrows,* facial nerve; *arrowheads,* hypoglossal nerve held by a rubber loop; *white C,* cerebellum; *black C,* internal carotid artery; *E,* turned-up ear. **c** After the dura has been closed and the sternocleidomastoid muscle, severed at the clavicle, has been placed onto the defect. *d,* Lyophilized dura; the sternocleidomastoid muscle is held with a pair of forceps. The facial and hypoglossal nerves have already been partly embedded in the muscle

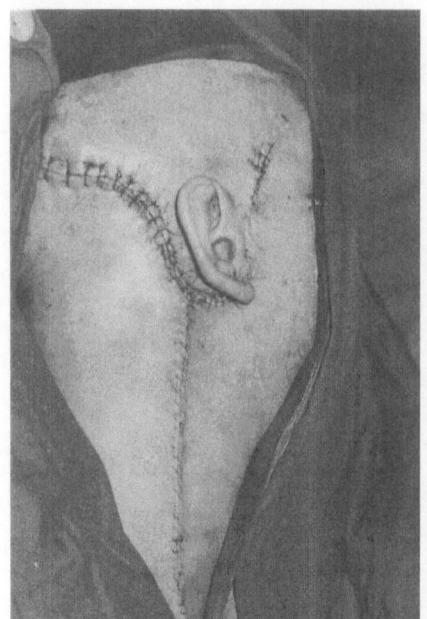

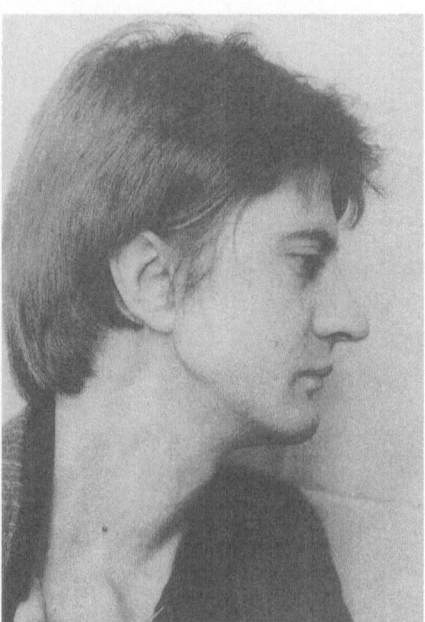

Fig. 5.11. a Y skin incision after surgery. **b, c** Satisfactory cosmetic result 2 years surgery. Slight paresis of the facial nerve still persists

fasciculi by the tumor, which made it impossible to preserve. Anatomical and functional preservation of the facial, glossopharyngeal, and hypoglossal nerves was successful. The entire petrosal segment of the internal carotid artery was exposed and detached from the tumor by extensive petrosectomy. The tumor – larger than a fist – was removed along with the dura of the middle and posterior cranial fossae. The large dural defect was sealed off watertight with lyophilized dura. The osseous defect left by petrosectomy was covered with temporal muscle and sternocleidomastoid muscle severed and turned up from the clavicule. The cosmetic result of the operation 1 year after surgery was very satisfactory (Fig. 5.11). Apart from the preoperative defects, additional transitory facial weakness and dysphagia occurred directly after surgery; however, improvement within a few weeks made operative correction unnecessary.

As extensive loss of blood can be expected in resection of such tumors, preoperative tumor embolization has important advantages. Viscosity-controlled embolization of the tumor nidus cannot be compared with "endovascular ligature" of feeding vessels. Although glomus jugulare tumors are basically supplied by vessels of the external carotid artery, significant collaterals from the vertebrobasilar system often represent a certain risk for preoperative embolization.

Frequently Encountered Lesions

The majority of tumors operated on via a laterobasal or infratemporal fossa approach are *glomus jugulare tumors* (Fisch 1978). These tumors originate in nonchromaffine glomera in the course of the tympanic nerve (ninth cranial nerve) and the auricular branch of the vagus nerve. They spread along the venous canals, first in the region of the jugular bulb and encroach on the posterior cranial fossa along the sigmoid sinus and on the apex of the petrous bone, the clivus, and the parasellar region via the pericarotid venous plexus. Due to their site of origin, they are supplied by the ascending pharyngeal artery. With regard to preoperative embolization it must be noted that there are almost no anastomoses in the region of the internal carotid artery, but numerous connections with the vertebral artery may exist via muscle branches. Due to partly intravascular growth of the tumor, the sigmoid sinus and the internal jugular vein usually have to be ligated and resected. Preservation of the ninth, tenth, and eleventh cranial nerves, which the jugular foramen transmits, is often impossible.

Carcinomas of the petrous bone are a further group of tumors warranting this approach. Infiltrated caudal cranial nerves often have to be sacrificed for the sake of radical excision. The extent of dural resection depends on local findings.

Advantages and Disadvantages of the Approach

The translabyrinthine approach is definitely too limited for lesions of the clivus or the apex of the pyramid; it is only sufficient when extended by a transcochlear approach (House and Hitselberger 1976) including displacement of the facial nerve. A precondition of the choice of this approach is that loss of hearing has already set in.

In exposure and displacement of the nerve, the danger of at least transitory facial paresis must be borne in mind. However, primary faciofacial anastomosis would be ideally suitable for cases of irreversible damage to the facial nerve by tumor infiltration, provided that the resected segment is not too long; otherwise, a nerve graft, such as greater auricular or sural nerve, should be used (Brackmann 1986) in order to obtain a nerve suture free from tension. If no appropriate proximal facial nerve stump is available, a different method, such as for example, a hypoglossal-facial anastomosis, has to be resorted to. We think that the functional results of such anastomoses are far better than those of accessory-facial anastomoses. A further option is the procedure published by Dott (1963), in which, as a second stage after suboccipital craniotomy, a nerve graft is placed between the stump of the facial nerve at the cerebellopontine angle and the extratemporal segment of the facial nerve.

Large processes not only involving the apex of the pyramid and the clivus, but reaching from the petrous bone to the infratemporal fossa, require an approach that is more extensive than the transcochlear approach. In the infratemporal approach described by Fisch (1977, 1978), the site of operation is extended into the infratemporal fossa after resection of the mandibular condyle (Fisch 1978 and case report above). Patients tolerate the resection of the mandibular condyle astonishingly well. A further disadvantage lies in the fact that the mandibular nerve has to be transected, which results in the subsequent deficit. Preservation of the caudal cranial nerves should be specially stressed. Compared with the facial and hypoglossal nerves, the recuperative capacity of the vagus and glossopharyngeal nerves is rather poor, even though they may have been preserved anatomically (Draf and Samii 1982; Kaye et al. 1984). In most cases, it is impossible to preserve the glossopharyngeal nerve when dissection into the infratemporal fossa is undertaken (Kumar and Fisch 1983). If slow tumor growth has already caused partial impairment of the vagus and glossopharyngeal nerves preoperatively, partial compensation will have set in and complete postoperative abolition is not dangerous. If permanent aggravated impairment of swallowing occurs postoperatively, redressment should be carried out as recommended by Denecke (1980). Isolated abolition of the glossopharyngeal nerve does not pose any essential problems; a vagus lesion, however, entails great danger of aspiration (Kumar and Fisch 1983).

After the jugular bulb has been exposed, direct lateral access is gained to the clivus so that the entire petrous segment of the internal carotid artery can be viewed as far as to the cavernous sinus. It is basically possible to extend the site of operation to the parasellar region as described by Fisch. This not only allows exposure but also mobilization of the entire petrosal segment, of the internal carotid artery, which leads to the possibility of reconstructing the carotid artery in case of tumor invasion (Jenkins and Fisch 1981; Glasscock et al. 1985). It must, however, be doubted whether petrosal aneurysms of the internal carotid artery, which occur very rarely, warrant the use of so extremely extensive approach as the one described (Glasscock 1983). The large defect left by this operation has to be covered. A pericranial flap, which is easier to prepare in cases of curved skin incisions than Y incisions, may be used to replace the dura. If one prefers not to use lyophilized dura, the exterior, fibrotic layer of fascia of the temporal muscle may be taken for patching. The deep wound cavity is packed with abdominal fat (Fisch 1978, Fisch and Pillsbury 1979) or temporal muscle and sternocleidomastoid muscle severed at the clavicule and turned up (Denecke 1966, 1969, 1978; Draf and Samii 1982). The latter procedure saves the patient a further, abdominal wound at the graft donor site.

As glomus jugulare tumors, to which this approach lends itself particularly well, are characterized by extensive vascularization, preoperative supraselective angiography is indispensable. At least at an initial stage, these tumors are exclusively supplied by the external carotid artery; preoperative embolization is therefore the appropriate method. New catheter technology renders embolization from the supply area of the vertebral artery possible, too. As embolization is not paramount to endovascular ligature of feeding vessels, but constitutes obliteration of the tumor nidus itself, if the proper embolization material has been chosen, it can be of decisive assistance for the operation (Lasjaunias 1981; Richling 1982; Hieshima et al. 1982).

We think that intraoperative teamwork between ENT surgeons and neurosurgeons is indispensable for this type of surgery, which is very time-consuming (Draf and Samii 1982; Classcock 1985). However, there have also been calls for a two-stage approach, with the dura forming the boundary separating the specialists (Kumar and Fisch 1983).

Summary: Laterobasal Approach

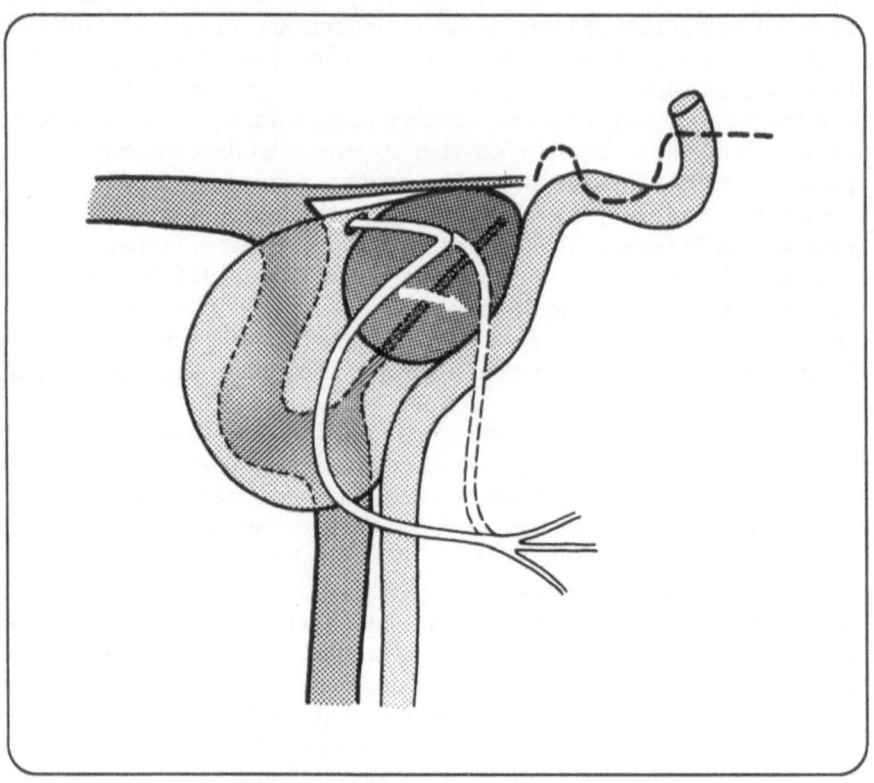

Indications: Extradural petroclival lesions
Lesions of the jugular foramen

Glomus Jugulare Tumors – Carcinomas – Chordomas

Advantages: Entire course of internal carotid artery under control
Feedings vessels are reached early
No temporal lobe or cerebellar retraction
Extension into middle and posterior cranial fossa possible

Disadvantages: Loss of hearing
(Transitory) impairment of seventh cranial nerve
Mandibular condyle resection
Mandibular nerve transection

Limits: Translabyrinthine: seventh cranial nerve
Transcochlear: internal carotid artery
Infratemporal: cavernous sinus (?)

Chapter 6: Temporal Approaches

Historical Survey

The frontotemporal approach is the most commonly used for the surgery of aneurysms in the anterior part of the circle of Willis as well as for sellar and parasellar lesions. After the Sylvian fissure has been split and dissection carried out along the mediobasal temporal lobe, structures behind the level of the dorsum sellae can be attained. The subtemporal approach as described by Drake (1965) allows direct access from laterally to structures behind the dorsum sellae. For sufficient access to the clivus and the structures of the prepontine space the following modifications have been proposed: large-scale extradural resection of the lesser wing of the sphenoid bone in the pterional approach (Yaşargil 1976), or the frontotemporal approach involving partial resection of the temporal lobe (Schisano and Tovi 1962; Symon 1982; Sekhar and Moller 1986). The subtemporal approach (Drake 1965, 1973; Peerless and Drake 1982) was extended toward the posterior fossa by splitting the tentorium (Schisano and Tovi 1962, case 2; Bonnal et al. 1964; Krayenbühl and Yaşargil 1975). A further interesting modification of the frontotemporal approach without resection of the temporal lobe was developed by Samii (1986), in which wide access to the posterior fossa from a frontotemporal direction is gained when the apex of the petrous bone is drilled off between the trigeminal impression and the internal auditory canal. Sugita et al. (1978) uses this modification for aneurysms of the basilar trunk. Kawase et al. (1985) developed a similar modification by fenestration of the tentorium to clip basilar artery aneurysms. Subtemporal craniotomy and additional temporary splitting of the zygomatic arch has been presented as an approach to basilar aneurysms (Pitelli et al. 1986) or tumors (Sekhar and Moller 1986; Kawase et al. 1987). The combination of a subtemporal approach and lateral suboccipital craniotomy is another option which will be presented in Chap. 9.

Anatomy

Anatomy Relevant to Surgery

The skin incisions are placed so that no damage is done to the frontal and ocular branches of the *facial nerve*. The frontal branch of the facial nerve runs to the forehead about 1 cm in front of the superficial temporal artery, so damage easily occurs in pterional craniotomy when dissection of the skin flap is carried out very superficially (Yaşargil 1984; Yaşargil et al. 1987).

The *nerves for the temporal muscle* branch off the mandibular nerve immediately below the foramen ovale and enter the muscle directly from the infratemporal fossa. This must be borne in mind in subtemporal approaches including a resection of the zygomatic arch (Knosp et al. 1991).

According to Yaşargil et al. (1975) and Yaşargil (1984a) and research done by us (Vorkapic et al. 1985), the *Sylvian fissure* should be split along the frontal lobe. Small bridging veins can be severed without sequelae. Arterial connections from the temporal to the frontal lobes via the Sylvian fissure are very rare.

The length of the *supraclinoidal internal carotid artery* of decisive importance in approaches to the basilar bifurcation and the clivus; it is 13.5 mm (8–18 mm) according to Lang (1981) or 19 mm (14–25 mm) according to Gibo et al. (1981). Dissection as far as the basilar bifurcation may be carried out either between the optic nerve and the internal carotid artery or lateral to the artery. This is where the posterior communicating artery runs dorso-medially; it may be an obstacle to dissection if it is large while the supraclinoidal segment of the internal carotid artery is short.

The number of branches from the *posterior communicating artery* is the same in large and in hypoplastic arteries (2–10 branches according to Yaşargil 1984; Saeki and Rhoton 1977). This is important if transection of the posterior communicating artery has to be considered (Yaşargil 1976, 1984b; Zeal and Rhoton 1978). In 12%–24% of all cases considered (Lang 1981; Yaşargil 1984) the posterior cerebral artery is of the fetal type, i.e., it originates from the internal carotid artery. In such cases, the P1 segment of the posterior cerebral artery is usually hypoplastic.

Diencephalic branches of the circle of Willis originate from the anterior cerebral artery (anterior inferior diencephalic branches) and from the posterior communicating artery (inferior diencephalic branches). Posterior inferior diencephalic branches (thalamoperforantes posteriores) from the posterior cerebral, the basilar, and the superior cerebellar arteries run through the posterior perforated substance into the posterior part of the thalamus. Of these, 80% branch off the precommunicating segment of the posterior cerebral artery, 14% from the basilar artery itself, and 5% from the superior cerebellar artery (Lang 1981, 1985). The "perforators" branch off the superior and dorsal aspects of the main vessel (Fig. 6.6) and, in exceptional cases only, the ventral aspect. No perforating arteries have ever been found on the anterior aspect of the basilar artery (Saeki and Rhoton 1977). Variations of origin were also described by Pedroza et al. (1986).

In some cases, the *posterior clinoid process* may obstruct the view of the basilar bifurcation (Fig. 6.3). The *veins of the temporal lobe* will be described in Chap. 9. The *superior petrosal sinus* runs along the ridge of the petrous bone and enters the most posterior part of the cavernous sinus above the porus of the trigeminal nerve. In 2.5% of all cases examined, the superior petrosal sinus divides around the trigeminal nerve (Boskovic et al. 1963). Usually, a large petrosal vein enters the superior petrosal sinus 5–7 mm laterally to the trigeminal nerve (Boskovic et al. 1963; Mitsushima et al. 1983). Only rarely does the petrosal vein drain caudally or mediocaudally to the trigeminal nerve into the superior petrosal sinus. At least during the embryonic stage, the superior petrosal sinus drains laterally into the sigmoid sinus (Knosp et al. 1987). The *blood supply to the tentorium* is by tentorial branches of the meningohypophyseal trunk and/or the inferior lateral trunk. The tentorial branch of the meningohypophyseal trunk runs along the trochlear nerve, which it also supplies. In rare instances this artery originates from the anastomosis between the middle meningeal and the ophthalmic arteries (Knosp et al. 1987). This variant is also referred to as the central meningeal artery (Lanz and Wachsmuth 1979) or artery of the free margin of the tentorium (Lasjaunias 1981).

The *greater superficial petrosal nerve* marks the way to the geniculate ganglion, which is only covered by thin bone, or may be situated immediately beneath the dura; according to Rhoton et al. (1967), this applies to 15% of all cases examined. (Paresis of the facial nerve in an extradural approach may occur by traction on the greater petrosal nerve.) The greater petrosal nerve roughly marks the course of the petrous segment of the internal carotid artery. Thus, the internal carotid artery can be found in the triangle formed by the middle meningeal artery, the foramen ovale, and the greater petrosal nerve. The course of the greater petrosal nerve is almost parallel to the ridge of the petrous bone. The middle meningeal artery and greater petrosal nerve are 8 mm apart (Parisier 1977; Lang 1981). The two petrosal nerves are located 2 mm apart (Lang 1981).

Table 6.1. Measurements and topographical relationships

Tentorial edge	to	IV nerve:	2 mm (3.6–1 mm)[a]
Tentorial edge	to	V nerve:	8 mm
IV nerve	to	V nerve:	6.6 mm (3.9–13.5 mm)[a]
V nerve	to	VII nerve:	7 mm (10.5–3.5 mm)[b]
Petrosal ridge	to	VII nerve:	4 mm (2.5–6.5 mm)[b]
Dorsum sellae	to	VI nerve:	19.3 mm (17.5–21.8 mm)[a]

IV nerve enters the dura: in 80% in the dorsal part of the *Wannenregion* ("trough")[c]
in 20% in the anterior petroclinoid fold

Free margin of the tentorium = ponto-mesencephalic sulcus
Floor of the middle cranial fossa = Frankfurt horizontal plane

Angle of petrosal ridge with median sagittal plane   128° +
Angle of carotid canal with median sagittal plane   124° +

[a] Ono et al. 1984
[b] v. Lanz and Wachsmuth 1979
[c] Lang 1981

The *tensor tympani muscle* with the Eustachian tube below it pass forward between the middle meningeal artery and the greater petrosal nerve (Paullus 1977).

Table 6.2. Topographical relationships

The *basilar bifurcation* is in:

51% of cases *at* the level of the dorsum sellae
30% of cases *above* the dorsum sellae
19% of cases *below* the dorsum sellae[a]

Distance between basilar bifurcation and dorsum sellae: 10 mms[b]

[a] Lang 1981
[b] Krayenbühl and Yaşargil 1979, in agreement with Samson et al. 1978

Key Steps of the Approach

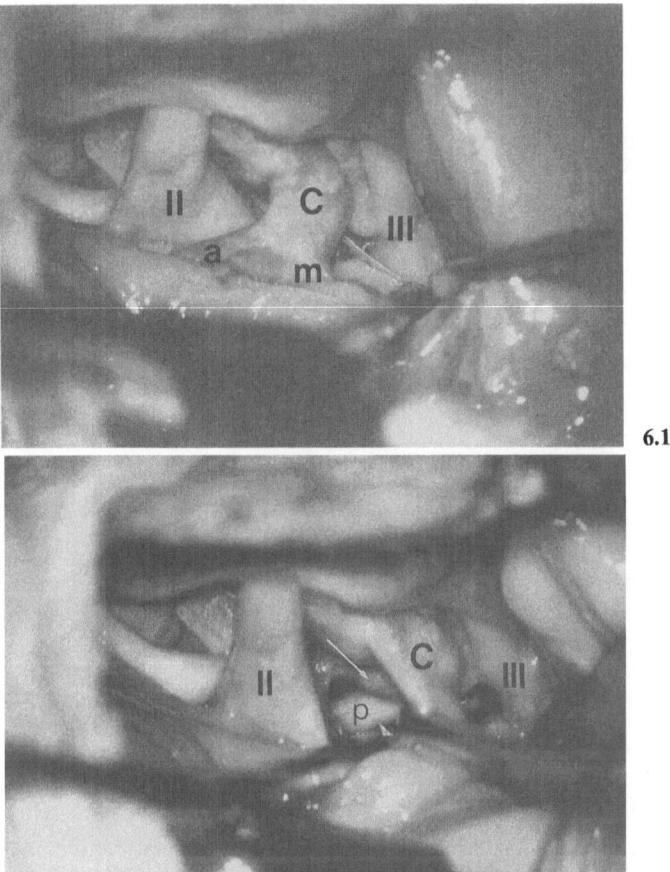

6.1

6.2

Fig. 6.1. Exposure of the optic nerve (*II*), the internal carotid artery (*C*), and the oculomotor nerve (*III*) after right frontotemporal craniotomy. The internal carotid artery bifurcates into middle cerebral artery (*m*) and anterior cerebral artery (*a*) quite early in its course, thus creating disadvantageous conditions for access to the basilar bifurcation

Fig. 6.2. When approaching between the optic nerve (*II*) and the internal carotid artery (*C*) on the right side (the latter is being held laterally with a dissector), there is only a little space for access to the prepontine space. Due to the deep bifurcation of the basilar artery, only the precommunicating segment of the posterior cerebral artery (*p*) can be exposed. The dorsum sellae (*white arrow*) bars the view of the basilar bifurcation

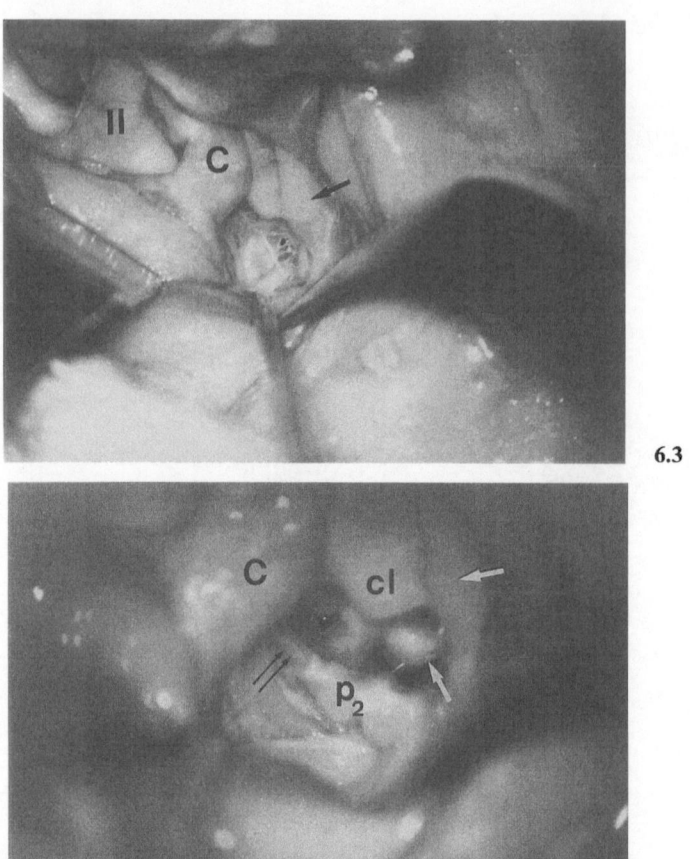

Fig. 6.3. The alternative approach to the internal carotid artery from laterally affords more space than the approach between the internal carotid artery and the optic nerve (cf. Fig. 6.1). After the proximal part of the Sylvian fissure has been split, the prepontine cistern is reached. *II,* Optic nerve

Fig. 6.4. After the arachnoid has been incised, the posterior cerebral arteries are followed in a proximal direction; now the posterior clinoid process (*cl*) obstructs the view of the basilar bifurcation. *C,* internal carotid artery; *black arrow,* oculomotor nerve; p_2, P2 segment of posterior cerebral artery; *white arrow,* superior cerebellar artery; *double arrows,* posterior communicating artery

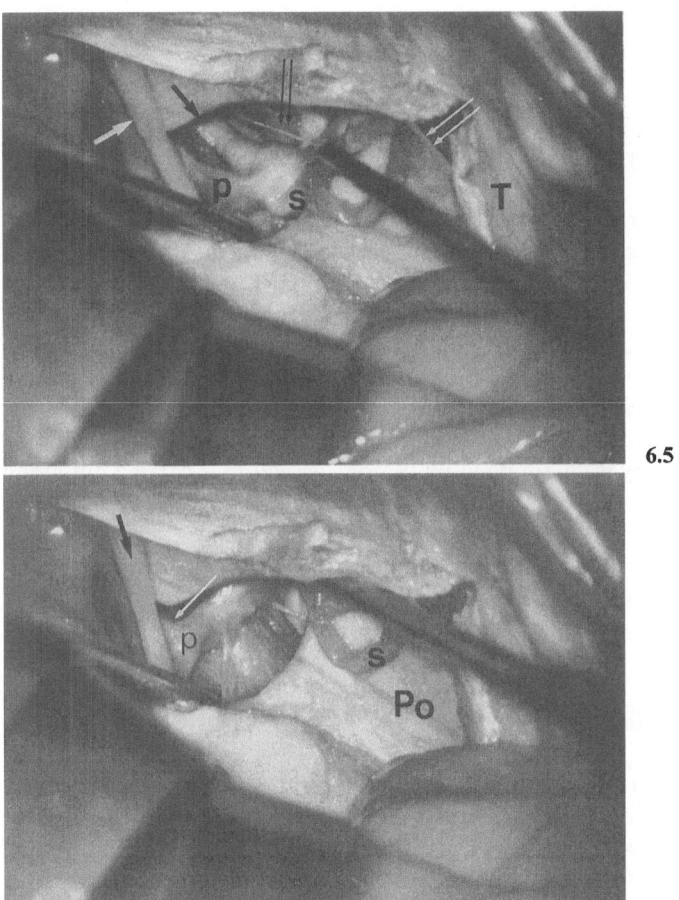

6.5

6.6

Fig. 6.5. As it is impossible to reach the basilar bifurcation without resecting the posterior clinoid process, this is achieved by via a subtemporal approach with incision of the tentorium (same specimen). The right (*p*) and left (*black arrow*) posterior cerebral arteries and right (*s*) and left (*black double arrows*) superior cerebellar arteries are visible. The basilar artery is slightly depressed with a dissector. *White arrow*, oculomotor nerve; *white double arrows*, trochlear nerve; *I*, tentorium

Fig. 6.6. Specimen showing the posterior inferior perforating arteries branching off the precommunicating segment of the posterior cerebral artery (*p*). They all exit at the dorsal aspect of the vessel. *White arrow*, point of entry of the posterior communicating artery; *arrow*, oculomotor nerve; *s*, superior cerebellar artery on the right; *Po*, pons

Surgery

Technique

Pterional Craniotomy

The skin incision starts in front of the ear close above the zygomatic arch and continues forward as far the hairline on the forehead (Yaşargil et al. 1987). After craniotomy, the lesser wing of the sphenoid bone is resected extradurally, which along with the splitting of the Sylvian fissure allows dissection between the frontal and temporal lobes. If there are plans to dissect below the anterior temporal lobe, craniectomy must be extended to the floor of the middle cranial fossa. After dural incision, the Sylvian fissure is first split basally and then the internal carotid artery is exposed during continuous CSF aspiration. If the internal carotid artery is surrounded by tumor, dissection is carried out along the middle cerebral artery. Extensive splitting of the Sylvian fissure should be avoided in the dominant hemisphere, if possible. Tumor resection is started in the middle cranial fossa, with the third and fourth cranial nerves mostly being extended over the tumor laterally and cranially. Thus, the tentorial notch and the petrous ridge are attained by removal of the tumor.

Subtemporal Craniotomy

The skin incision starts at the middle of the zygomatic arch and leads around the ear in a semicircle. The zygomatic process marks the level of the floor of the middle cranial fossa, so the skin incision has to be large enough to reach it, in order to ensure that the base is attained. Craniotomy is centered directly above the ear, with the posterior burr hole just above the point where the suture of the temporal squama and the lambdoid suture join. In this way, damage to the transverse sinus is avoided. The upper rim of the sinus is situated 2 cm above the Frankfurt horizontal plane (Seeger 1980); Lang (1981) reported dimensions of 5–10 mm. To avoid retractor pressure on the temporal lobe, craniotomy must be extended to the floor of the middle cranial fossa. The basal dura is opened and fixed basally by staying sutures. Minor bridging veins of the temporal lobe may be severed, but under no circumstances must Labbé's vein, which ends in the transverse sinus, be transected. The further procedure now depends on how far the tentorium, cavernous sinus, and the clivus are involved in the tumor. The first step is splitting the tentorium as far as the petrous ridge, the second assessing how far the cavernous sinus is involved, and the third step is establishing tumor extension on the clivus.

Incision of the Tentorium

First of all, the position of the trochlear nerve is of importance for the incision of the tentorium (Samii and v. Wild 1981). When the mediolateral temporal lobe is lifted, the trochlear nerve may be drawn upward behind the tentorial edge without the arachnoid of the ambient cistern being opened; then the tentorium may be incised on its free margin. Only then may the cisterna be opened. When the tentorium is transected, the incision runs a few millimeters off the edge of the petrous bone so as to avoid opening the superior petrosal sinus.

Tumor Involvement of the Cavernous Sinus

If the tumor has invaded the cavernous sinus, the further procedure depends on the function of the second through sixth cranial nerves and the surgeon's experience. If operation in the region of the cavernous sinus is decided on while the cranial nerves are intact, the sinus is opened in what is called Parkinson's triangle (Fig. 6.7).

The dura is incised between the trochlear nerve and the first branch of the trigeminal nerve. If the tumor has completely invaded the cavernous sinus, there is no danger of venous hemorrhage from the sinus, as it is filled with tumor tissue. The cranial nerves are exposed at optimum magnification and the surrounding tumor tissue is removed in a piecemeal manner. We consider the abducent nerve at most risk, since its relation to the lateral cavernous sinus wall is not as clear-cut as that of the third, fourth, and fifth cranial nerves. Of course, in this region special attention must be paid to the internal carotid artery, which may pass forward on a rather unusual course due to tumor compression and encasement. Even though preoperative angiography may reveal the type of displacement, the intraoperative findings regarding whether the tumor may be detached from the vessel or not are of decisive importance. The surgeon is guided first and foremost by the pulsations of the vessel. The risk of direct damage to the internal carotid artery during dissection is not as great as the risk of indirect damage by tearing out a vessel supplying the tumor. Doppler sonography has proven an efficient intraoperative aid as it traces the course of vessels and measures the distance at which they are located. Increasing experience in tumor surgery in the cavernous sinus region (Parkinson 1965, 1973; Dolenc 1983, 1985; Dolenc et al. 1987b, Sekhar and Moller 1986; Sekhar and Samii 1986; Kawase et al. 1987; Perneczky et al. 1987) has shown that tumor removal is possible while preserving the function of the cranial nerves.

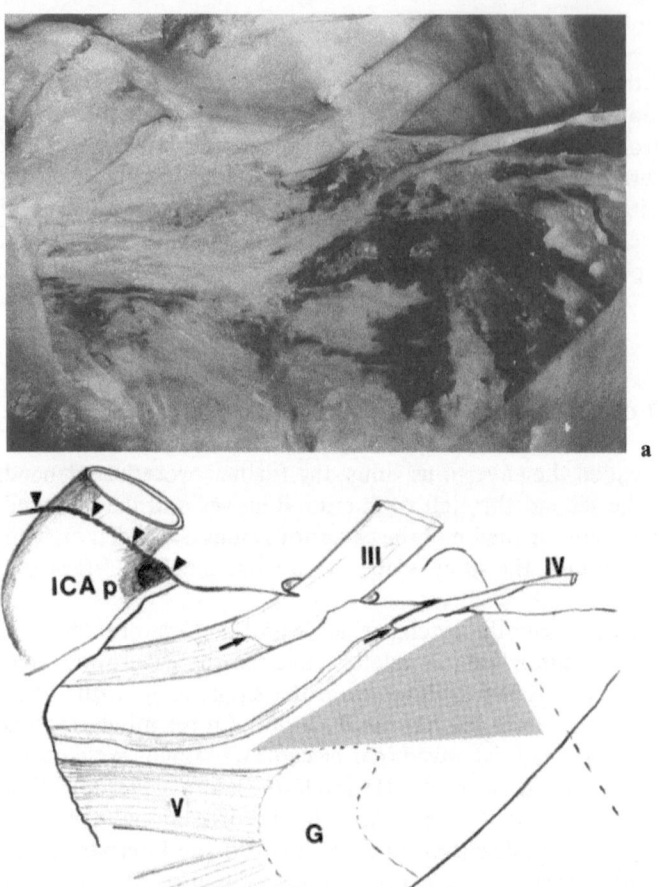

Fig. 6.7 a, b. Lateral wall of the cavernous sinus and Parkinson's triangle: the third and fourth cranial nerves mark the cranial limits, V_1 (first division of the trigeminal nerve) and/or the trigeminal ganglion the caudal limit; the clivus forms the dorsal limit. **a** Anatomical specimen of cavernous sinus, venous filling with colored latex milk; **b** dissection diagram. The inner dural layer of the lateral wall of the cavernous sinus has been removed. The third, fourth, and fifth cranial nerves (III–V) are interwoven with the outer dural layer. Dissection of the cranial nerves in their epidural segments far more difficult than in their cisternal segments as they lack the surrounding arachnoidal sheath. At their dural pori, they are surrounded by an arachnoidal pouch for a short distance (*arrows*). In the area Parkinson's triangle the outer layer of the lateral wall of the cavernous sinus is missing, making the region appear a little more transparent (see also Umanski and Nathan 1982; Rhoton et al. 1979). The anterior clinoid process has been removed and the paraclinoidal segment of the internal carotid artery (*ICA p*) has been exposed. The fibrous ring around the internal carotid artery at the area of dural transition is marked (*arrow heads*)

Resection of the Apex of the Petrous Bone

(Modification according to Samii)

If it is impossible to reach major portions of the tumor in the posterior cranial fossa along the clivus by splitting the tentorium, we propose the following modification. Before starting resection of the apex, the foramen spinosum, middle meningeal artery, foramen ovale, and the course of the petrosal nerves must be located with certainty. The petrosal nerves are easier to recognize by their blood supply from the middle meningeal artery than to see through the dura. After excision of an appropriate flap of dura (Fig. 6.8), resection of the apex of the petrous bone may be embarked on. The dura is incised in semi-circular fashion to the porus of the trigeminal nerve (Fig. 6.8). Osseous exposure of the apex of the petrous bone follows. The apex is then drilled off with a diamond drill until the dura at the posterior fossa aspect of the petrous bone is reached. Any hemorrhage from the inferior petrosal sinus (located medially) is controlled by packing. The petrous bone may be resected as far as the upper rim of the internal auditory canal, so we are able to expose the seventh and eighth cranial nerves in this region and detach them from the tumor. When the petrous segment of the internal carotid artery is reached, we have also arrived at the caudal limit of bone resection. The approach to the posterior cranial fossa is thus defined by the trigeminal nerve, the internal carotid artery caudally, and the seventh and eighth cranial nerves laterally (Fig. 6.11).

This modification affords a sufficient view of the cerebellopontine angle and prepontine space (Sekhar and Samii 1986; Samii and Draf 1989). If serious preoperative damage to the trigeminal nerve or tumor infiltration of Meckel's

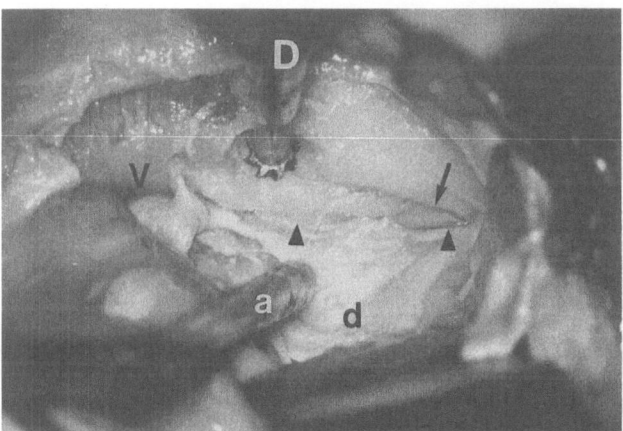

Fig. 6.8. After right frontotemporal craniotomy, the dura is incised in a semicircle, with the incision reaching into Meckel's cavity. The dura (*d*) is detached as far as the petrous ridge (*arrowheads*) and held in place with an aspirator (*a*). *V,* Trigeminal nerve; *arrow,* groove of the superior petrosal sinus; *D,* drill at the apex of the petrous bone

cavity are revealed intraoperatively, resection of the trigeminal nerve can open up an even wider space between the third and fourth cranial nerves on the one hand and the seventh and eighth cranial nerves on the other. If the central stump of the trigeminal nerve can be recognized, reconstruction can be performed (Samii 1991). This modification, together with displacement of the brain stem by the tumor, makes it possible to remove tumors reaching as far as the midline and to obtain an unusual view of the posterior cranial fossa (see case report below).

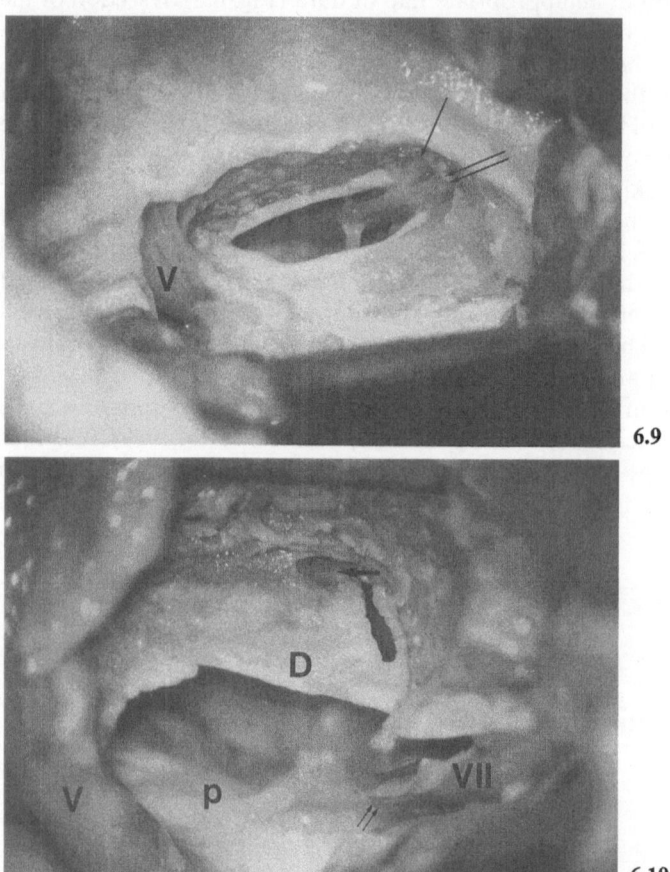

Fig. 6.9. After the apex of the petrous bone has been resected, the internal auditory canal is exposed. The facial nerve (*arrow*) is separated from the superior vestibular nerve (*double arrow*) by the vertical crest. *V*, Trigeminal nerve

Fig. 6.10. Situation after the apex of the petrous bone has been further drilled off between the trigeminal nerve (*V*) and the seventh and eighth cranial nerves (*VII*). The dura at the posterior aspect of the petrous bone (*D*) has been exposed. The inferior petrosal sinus has been opened (*arrow*): view into the prepontine space, pons (*p*) and loop of the anterior inferior cerebellar artery in the internal auditory canal (*double arrow*)

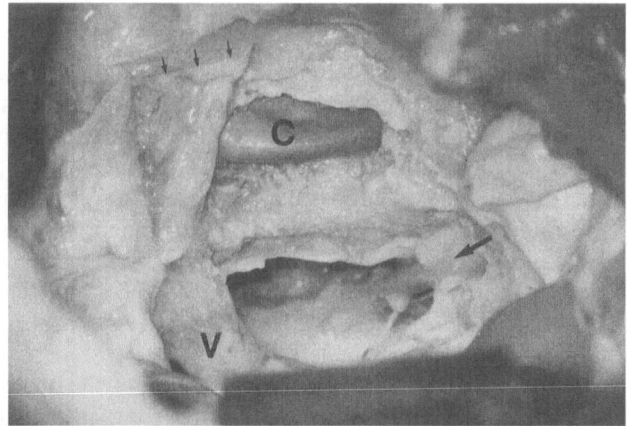

6.11

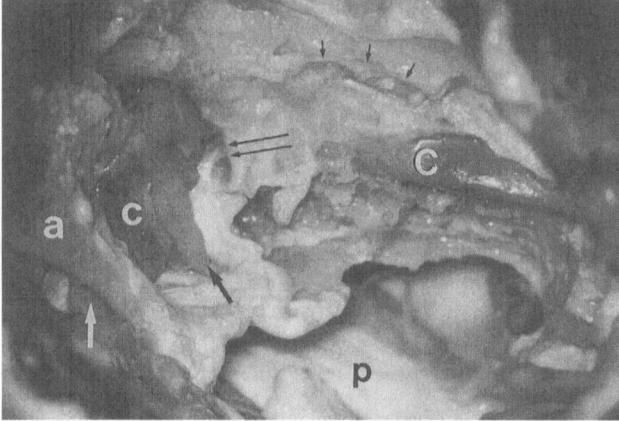

6.12

Fig. 6.11. The petrous segment of the internal carotid artery (*C*) has been exposed between the trigeminal nerve (*V*) and the internal auditory canal (*arrow*). The foramen ovale is marked by *small arrows*

Fig. 6.12. After the trigeminal nerve has been "sacrificed," the most medial portion of the apex of the petrous bone and the adjacent clivus are further resected. The cavernous sinus has been opened in its posterior part. The sixth cranial nerve (*black arrow*) is in contact with the intracavernous segment of the internal carotid artery (*white*) on its lateral side. *Black C*, Internal carotid artery, petrous segment; *a*, anterior petroclinoid fold; *white arrow*, oculomotor nerve at its point of entry into the dura; *double arrow*, rim of the inferior sphenopetrosal ligament; *small arrows*, foramen ovale; *p*, pons

Case Report

For years this man had been suffering from amaurosis and ophthalmoplegia as well as trigeminal nerve palsy on the left side and marked visual impairment on the right. Plain films showed complete absence of the sella and an extremely flat sphenoid ridge (Fig. 6.13). CT scanning revealed a meningioma extending over the entire skull base (Fig. 6.14). It is impossible to excise a tumor of this size in one operation. One could see it as consisting of the following three parts: one portion located in the middle and posterior cranial fossa, a meningioma of the tuberculum sellae, and an intracavernous portion on the right. The tumor was operated on according to these imaginary subdivisions, with surgery starting from the large portion on the left.

Because of complete disfunction of the optic nerve and palsy of the oculomotor nerves and the left trigeminal nerve, the modified frontotemporal approach described above lent itself well to the first stage of surgery. Thus, resection of the tumor portion in the middle cranial fossa and – after the apex of the petrous bone had been drilled off – of the entire tumor portion in the posterior cranial fossa succeeded. Subsequently, the optic nerves and the internal carotid arteries on both sides as well as the basilar artery with its branches and the contralateral oculomotor nerve were exposed (Fig. 6.17). The exact scope of the petrous bone resection can be clearly seen in the postoperative CT scan (Fig. 6.16). Upon exposure of the intrapetrous segment of the internal carotid artery, evulsion of a tumor vessel caused arterial hemorrhage which had to be controlled by compression. This was the site of origin of an iatrogenic aneurysm of the internal carotid artery which was discovered in the second operation.

Due to progressive visual impairment on the right side the next step to be planned was the removal of the tumor portion on the right. The cavernous sinus turned out to be entirely invaded by the tumor. Although the oculomotor nerve had been detached from the tumor intradurally, it was decided not to open the cavernous sinus in order to preserve the functioning oculomotor nerve of the remaining healthy eye. The optic canal was decompressed and the supra- and endosellar portions of the tumor were resected in a third operation. The postoperative radiological result was satisfactory; however, progressive visual deterioration of the patient and subsequent blindness could not be stopped.

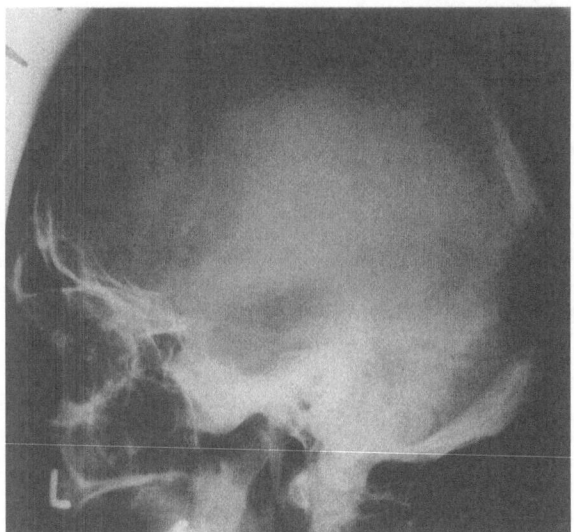

Fig. 6.13. Plain film showing destruction of lesser wing of the sphenoid bone and sella as well as the clivus

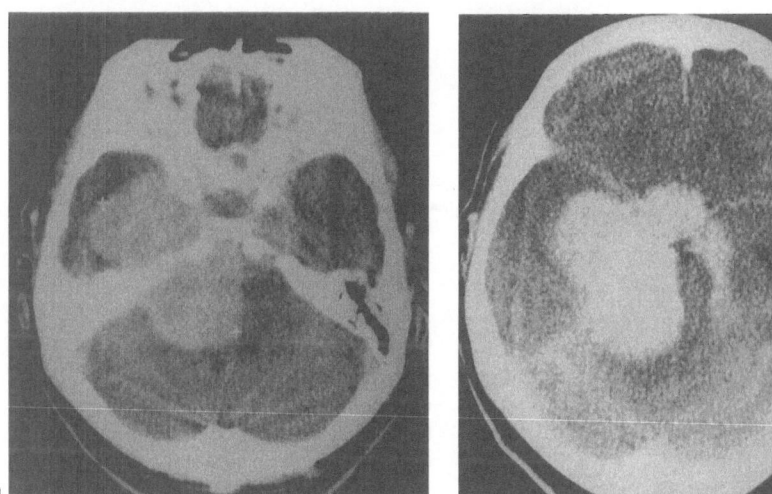

a

b

Fig. 6.14a, b. Preoperative CT scans revealing a basal meningioma extending a considerable way into the middle and posterior cranial fossa (**a**). The tumor has encroached on the right cavernous sinus and there is a significant space-occupying suprasellar lesion (**b**)

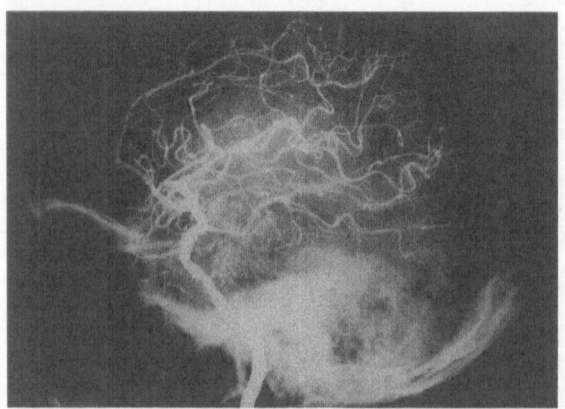

a

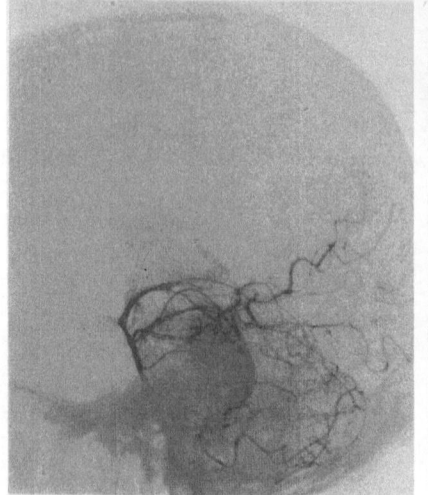

b

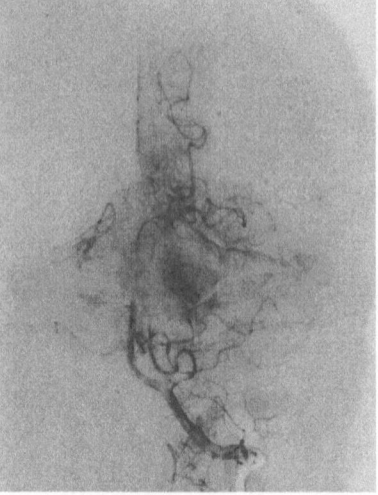

▲

Fig. 6.15. a Angiography of the left internal carotid artery. The internal carotid artery has been displaced in its intrapetrous and intracavernous segments. In addition, the posterior cerebral artery has been displaced upwards. Angiography of the vertebral artery reveals the extent of the prepontine space-occupying lesion

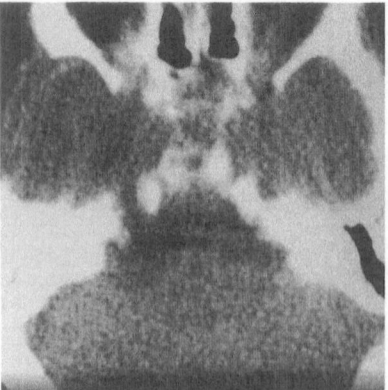

Fig. 6.16. CT after the third operation

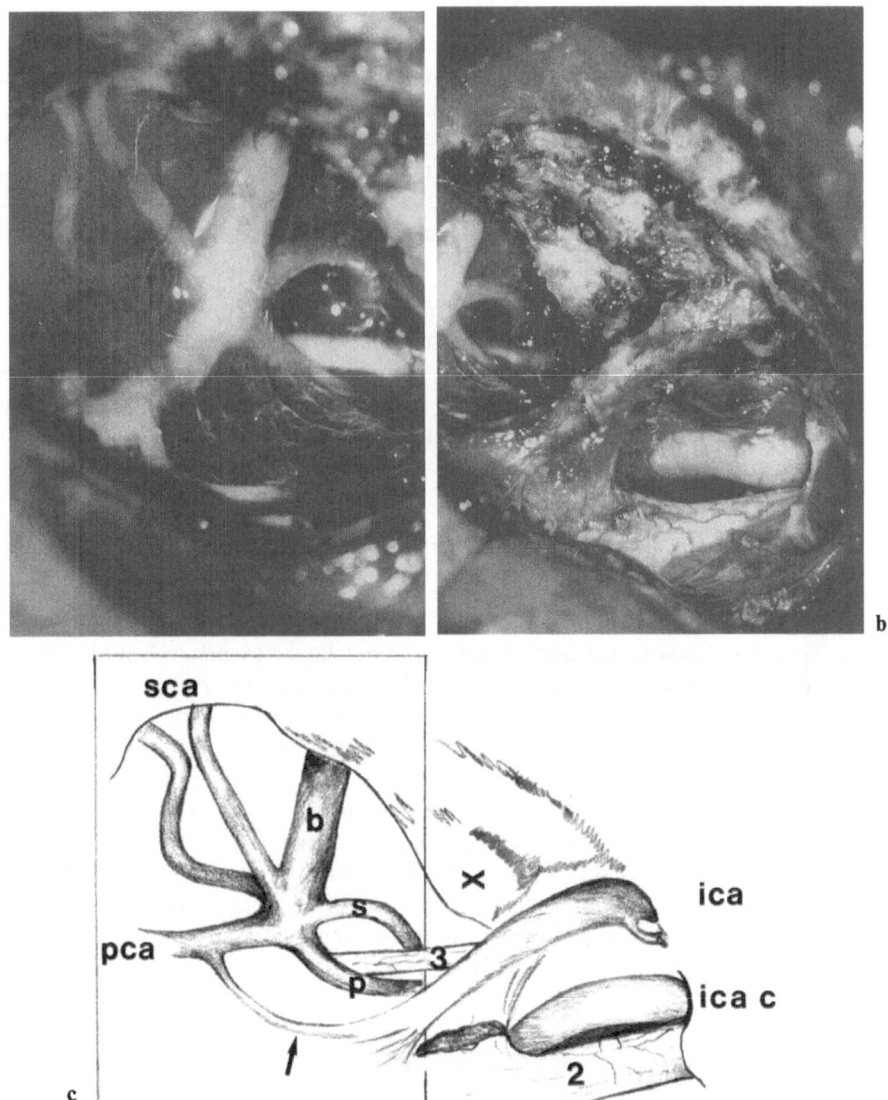

Fig. 6.17. a Intraoperative site after tumor removal by means of the modified frontotemporal approach (first operation). Drilling of the apex of the petrous bone affords an unusual view of the posterior cranial fossa. The detail (**b**) shows the basilar bifurcation (duplication of the ipsilateral superior cerebellar artery) and the contralateral optic and oculomotor nerves. **c** Diagram of the operation site. *b,* Basilar artery; *sca,* double superior cerebellar artery on the left *pca,* left posterior cerebellar artery; *p,* right posterior cerebellar artery; *s,* right superior cerebellar artery; *3,* right oculomotor nerve; *ica,* left internal carotid artery; *ica c,* contralateral internal carotid artery. The *arrow* marks the course of the left posterior communicating artery. *2,* Right optic nerve; *X,* resected apex of the petrous bone

Frequently Encountered Lesions

The majority of tumors are *meningiomas* with major portions located in the middle cranial fossa, encroaching on the petrous bone and the clivus. In such cases the third and fourth cranial nerves are displaced laterally or upward by the tumor. The seventh and eighth cranial nerves are displaced laterally. It is often very difficult to find the trigeminal nerve, the fibers of which are split up, especially if the tumor has invaded Meckel's cavity. Meningiomas grow along the areas of entry of cranial nerves and veins; thus, it comes as no surprise that involvement of the cavernous sinus is fairly common.

In many cases *chordomas, chondromas,* and *chondrosarcomas* are not limited to the clivus – or not to the clivus alone – but also encroach on the apex of the petrous bone and the middle cranial fossa, which means that they can be reached via a temporal approach. This is especially important in patients in whom hearing has been preserved and the laterobasal approach would entail too much of a sacrifice in the form of hearing loss (cf. Chap. 5).

Cases in which *neurinomas* of the trigeminal nerve erode the apex of the petrous bone and reach the clivus are rare.

Another indication for the use of this approach is given by *aneurysms of the basilar artery*; that is not only those of the basilar bifurcation, but also those of the basilar trunk, the superior cerebellar artery or the proximal posterior cerebral artery, which are all rather uncommon. Aneurysms of the basilar bifurcation represent 5% of all intracranial aneurysms, and this is thus the most common aneurysm site (50%) in the posterior circulation (Yaşargil 1984). Apart from the different sizes of aneurysm or their fusiform shape, which often renders surgery impossible, the projection of the dome of the aneurysm is of decisive importance. Aneurysms projecting ventrally or cranially do not usually involve the "perforators" as often as those projecting posteriorly. Operative results in posteriorly projecting aneurysms are considerably worse (Yaşargil 1984; Drake 1973).

Advantages and Disadvantages of the Approach

Since it makes a great difference in the selection of an approach whether a pathological process is a space-occupying lesion or a vascular malformation, we will discuss these separately.

Space-occupying lesions cause anatomical changes as they displace parts of the brain, cranial nerves, and vessels. This has to be taken as much into account as the advantage that considerable space becomes available after piecemeal tumor resection. In many cases, complete or partial damage of cranial nerves has already occurred preoperatively, so one may also calculate that they may be "sacrificed" if recovery is unlikely. These circumstances make frontotemporal approaches more useful for tumors of the clivus.

Such "favorable" circumstances are usually absent in vascular malformations, however, such as basilar aneurysms: For these the approach is mostly determined by the "normal" anatomical conditions, and cranial nerve lesions are the exception. The frontotemporal (pterional) approach can only be used on tumors of the clivus which also extend same way into the middle cranial fossa. In such cases the third and fourth cranial nerves are displaced cranially and the vessels of the Sylvian fissure displaced upward. This enables the surgeon to make optimum use firstly of the advantages created by splitting the Sylvian fissure, secondly of the space gained by removing the tumor in a piecemeal manner. The posterior cranial fossa is accessible after the tentorium has been split and – if necessary – the apex of the petrous bone has been resected. This improves the view caudally and, to a somewhat lesser degree, medially. The limiting factors here are the posterior communicating artery and the dorsum sellae; thus, tumor portions behind or below the dorsum sellae are difficult (or even impossible) to reach. In this context, Symon (1982) and Sekhar and Moller (1986) recommended resection of the temporal pole to improve the angle of approach to medial structures. Often this resection merely anticipates removal of brain damaged by forced retractor pressure, something that should be avoided. In order to reduce the pressure on the temporal lobe one has to design approaches which include resection of parts of the skull base rather than forced brain retraction (Al Mefty 1989; Knosp et al. 1991). The pterional approach as shown by Yaşargil includes extradural resection of the lesser wing of the sphenoid bone and was one of the first of that type (Yaşargil 1984b; Yaşargil et al. 1975, 1976).

Especially in lesions that involve the posterior aspect of the cavernous sinus, temporary resection of the frontal process of the zygomatic bone and the lateral rim of the orbit may result in a decisive improvement of view (Dolenc et al. 1987a; Perneczky et al. 1988; Al Mefty 1987). The frontal branches of the facial nerve become more endangered the further one proceeds into the skull base. Another option is to temporarily transect the zygomatic arch, which makes it possible to turn the temporal muscle more caudally in order to reach the tentorial notch below the anterior temporal lobe. However, the main use of this approach is for lesions in the region of the cavernous sinus (Sekhar and Moller 1986; Kawase et al. 1987). When mobilizing the temporal muscle in this way, attention must be paid to its motor nerves so that no cause muscular atrophy is caused (Pitelli et al. 1986; Knosp et al. 1991). These branch from the mandibular nerve directly to the temporal muscle and are very short.

An entirely subtemporal approach must be chosen for the approach of lesions situated behind the dorsum sellae and extending beyond the midline. The angle of approach is mainly from laterocaudal to mediocranial, i.e., although the tentorium has been split there is a caudal limitation of the approach into the posterior cranial fossa. Resection of the skull base, which allows different options in frontotemporal approaches, is of limited use at the petrous bone due to the tympanic cavity. Craniotomy may, however, be extended by a lateral suboccipital approach (see Chap. 7) to improve access to the posterior fossa.

The frontotemporal approach (or one of its modifications) therefore lends itself well to tumors occupying considerable space in the middle cranial fossa whose extension at the clivus is mainly paramedian. For tumors occupying little space in the middle cranial fossa, but extending beyond the midline at the clivus, a classical subtemporal approach must be chosen. For processes which reach beyond the midline and a long way caudally one will have to use a combined supra-infratentorial approach, in particular in cases in which the tentorium has been invaded. One of the disadvantages of frontotemporal approaches including extensive splitting of the Sylvian fissure is the risk of hemiparesis: splitting as far as to the periphery should therefore be avoided in the dominant hemisphere. Moreover, there is danger of brain damage by retraction of the temporal pole and the mediobasal temporal lobe. Resection of the anterior third of the temporal lobe – as has been recommended (Symon 1982; Sekhard and Moller 1986) – must be considered in the context of one's personal surgical technique and experience. It is often constitutes no more than an anticipation of the removal of malacic brain damaged by retraction. The drawback of all subtemporal approaches is the problems caused by the temporal lobe, whether intraoperatively by difficulties in relation to veins or postoperatively by swelling, intracerebral hemorrhage, or the development of epilepsy.

Essentially, there are two approaches up for comparison regarding aneurysms of the basilar bifurcation: the subtemporal approach recommended by Drake (1965) and the pterional approach of Yaşargil et al. (1976). The advantages and disadvantages of each depend not only on the surgeon's preference, but also on anatomical and pathoanatomical conditions (Solomon and Stein 1988). In discussing the choice of approach, the size of the aneurysm is as much of importance as the location of the basilar bifurcation in relation to the dorsum sellae, the direction in which the aneurysm dome projects, and the course of the perforating vessels.

Large space-occupying or complex aneurysms are almost impossible to reach via a pterional approach (Symon 1982), let alone to clip safely; in such cases, the method of choice ought to be the subtemporal or a combined supra-infratentorial approach (Symon 1982). The basilar bifurcation cannot be reached by means of a pterional approach if it is situated too far beneath or high above the level of the dorsum sellae; consequently, this is another indication for the subtemporal approach (Samson et al. 1978). Furthermore, it is difficult, if not impossible, to gain a good view of the perforators via a pterional approach if the aneurysm dome projects dorsally, whereas a dome projecting ventrally – toward the surgeon and away from the perforators – is easier to survey and to separate from the perforators (Yaşargil 1984).

The main advantage of the pterional approach is the simultaneous view it allows of both posterior cerebral arteries and the brain stem vessels originating from them: by contrast in a subtemporal approach coming from laterally mobilization of the aneurysm is required to enable the surgeon to see vessels from the origin of the contralateral posterior cerebral artery first, which increases the danger of aneurysm rupture. Dissection as far as the basilar bifurcation by the pterional approach is carried out either between the optic nerve and the internal carotid artery or lateral to the artery; an alternative would be above the carotid bifurcation (Sugita 1985) or medial to the posterior communicating artery between the perforating diencephalic branches (Sugita et al. 1979) (inferior diencephalic branches).

Thus, the approach basically depends on the length of the internal carotid artery before it divides – considerable difficulties may arise from a short supraclinoidal segment. If the posterior communicating artery is sturdy and turns laterally, or if a fetal type of posterior cerebral artery is present, access to the basilar artery may be rendered impossible. Severing of the posterior communicating artery has been described in several cases, with none too unfavorable results (Yaşargil 1984). In such cases, Sugita (1985; Sugita et al. 1979), however, recommends an approach between the perforating vessels of the posterior communicating artery rather than the severing of the posterior communicating artery, even though the artery be of relatively small caliber. The posterior clinoid process (Fig. 6.4b) may prove a further obstacle on the way to the basilar bifurcation (Tulleken and Luiten 1986); resection may be necessary (Yaşargil 1984).

None of these problems centered around the pitfalls of the approach exist in the subtemporal approach. However, considerable difficulties arise around the issues of causing damage to the temporal lobe or the insufficient view of the contralateral posterior cerebral artery together with the contralateral perforating vessels (see above), which is another reason why some surgeons prefer the pterional approach (Sugita et al. 1979). The more the aneurysm projects posteriorly (or posteroinferiorly), the more favorable the subtemporal approach is, because then in a pterional approach the neck of the aneurysm would be hidden behind the basilar artery.

Another obstacle on the way to the basilar bifurcation in a subtemporal approach is the oculomotor nerve, which traverses the prepontine cistern between the posterior cerebral artery and the superior cerebral artery, and crosses the field of operation (Figs. 6.5, 6.6, 9.4)

Summary: Temporal Approaches

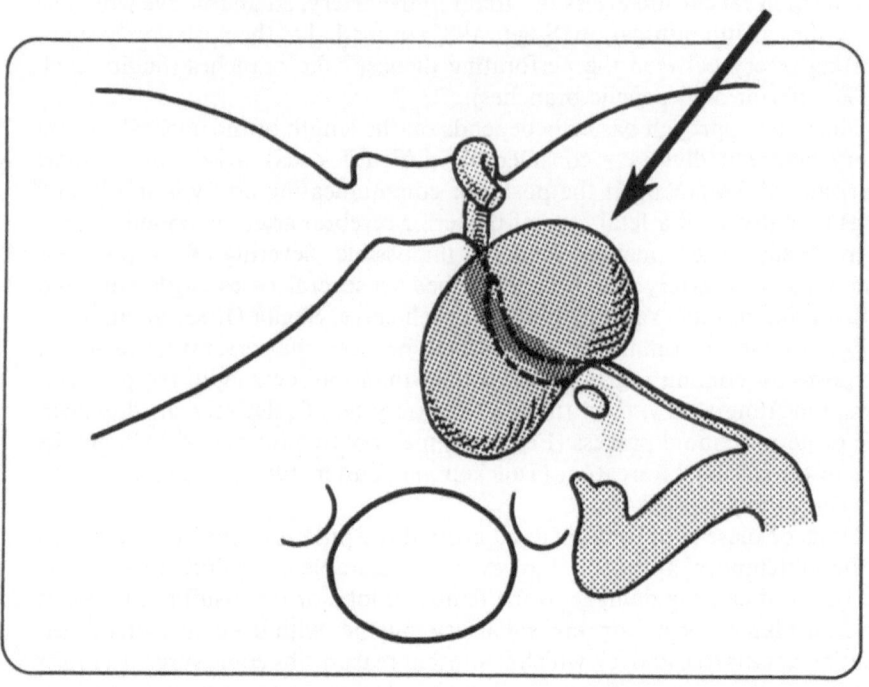

Indications: Petroclival tumors with mainly middle cranial fossa
extension

Meningiomas – Chordomas – Basilar aneurysms

Advantages: Access to the middle and posterior cranial fossa in one stage
Modifications tailored to the lesion are possible

Disadvantages: "Temporal lobe problems"
Limited access to the posterior fossa

Limits: Lower clivus
Midline

Chapter 7: Lateral Suboccipital Approach

Historical Survey

In the pioneering days of neurosurgery, lesions in the clival region were still considered inoperable (Cushing and Eisenhardt 1938; Castellano and Ruggiero 1953; Olivecrona 1967). The few cases in which tumor extirpations were carried out by Cushing in 1914 and Olivecrona in 1927 did not end in success, which comes as no surprise considering that mortality in cases of acoustic neurinoma was far more than 10% (Krause 1904; Cushing 1917; Dandy 1917, 1941). Improved diagnostic techniques and the development of more sophisticated surgical techniques led to Dandy's unilateral suboccipital approach to the cerebellopontine angle (Dandy 1925, 1941; Dandy et al. 1963). However, despite progress in this field, tumors of the clivus continued to be regarded as inoperable (Castellano and Ruggiero 1953; Olivecrona 1967), with very few exceptions (Dechaume and Wertheimer 1936). The introduction of microsurgical techniques and, as a consequence, better results in acoustic neurinoma surgery (Rand and Kurze 1965; Malis 1975; Yaşargil 1976, Yaşargil et al. 1977, 1980; Koos and Perneczky 1982, 1985; Koos et al. 1976, 1985; Rand and di Tullio 1985; Samii 1981, 1985) made it possible to reach petroclival and clival tumors by means of a lateral suboccipital approach.

We think that the lateral suboccipital approach as described for the removal of acoustic neurinomas is the most appropriate way of removing intradural petroclival lesions. A slight change in positioning also makes it very well suited to intradural lesions at the anterior margin of the foramen magnum. We will describe this modification separately below, as we find it important to point out certain details. A separate chapter (Chap. 9) will be dedicated to a combination of the lateral suboccipital approach and a posterior subtemporal approach.

Anatomy

Anatomy Relevant to Surgery

The *jugular foramen* is the main route for drainage of cerebral blood. The sigmoid sinus terminates in the posterior lateral part (pars vasculosa) of the foramen, while the inferior petrosal sinus passes through the anterior part and occasionally only joins the jugular vein extracranially (v. Lanz and Wachsmuth 1979). The caudal cranial nerves (ninth, tenth and eleventh) also end in the anterior part. The glossopharyngeal nerve has a dural porus of its own, whereas the tenth and eleventh cranial nerves share one dural porus (Fig. 7.4 and 9.1 a). The jugular foramen is divided in two parts by a more or less marked jugular process of the petrous bone; only very rarely is there a complete subdivision (Lang 1981, 1986). In rare instances, the bulb of the jugular vein reaches as far as the internal auditory canal.

The *sigmoid sinus* is situated lateral to the occipitomastoid suture. In trepanations, the mastoid emissary marks the point at which one has arrived above the sigmoid sinus. There are considerable differences in the dimensions of the right and left sigmoid sinuses the one on the right being larger, which is explained by its facilitation of the backflow of blood to the right heart. Therefore, the large superior sagittal sinus mainly drains to the right (Seeger 1980; Bisaria 1985).

The *petrosal vein* is formed from the anterolateral cerebellar veins, the pontomedullary vein, and several smaller veins. It usually drains into the superior petrosal sinus lateral to the trigeminal nerve, in rare cases laterocaudal to it (Fig. 7.3). The superior petrosal sinus becomes larger laterally, and the connection of the superior petrosal sinus to the posterior part of the cavernous sinus is usually not well developed; this supports the view that drainage mainly occurs via the sigmoid sinus, which accords with research done on fetuses (Knosp et al. 1987).

Veins of the *hypoglossal canal* pass the basis of the occipital condyle together with the twelfth cranial nerve, which occasionally is double (see Figs. 2.4 c, 9.2 a). These veins were connected to the jugular foramen in one fourth of the cases examined, to the inferior petrosal sinus in one-third, and to the marginal plexus of the foramen magnum in one-third (Lang 1981). An *oblique occipital sinus* is a rare variety which occurs frequently only in children; in such cases this sinus crosses the occipital squama and extends to the jugular foramen at the margin of the foramen magnum.

Key and Retzius were the first in 1875, to describe *cisterns* as extensions of the basal subarachnoidal space. Microsurgical techniques, however required a more detailed description of the basal cisterns (Yaşargil 1984; Yaşargil et al. 1976). The compartmentalization in the basal cistern system can be particularly well observed in the lateral suboccipital approach. There is a lateral cerebellomedullary cistern containing the ninth, tenth, and eleventh cranial nerves, the vertebral artery and the proximal inferior posterior cerebellar artery. The superior cerebellopontine cistern adjoins this cranially; it conveys the seventh and eighth cranial nerves and the anterior inferior cerebellar artery and is connected to the arachnoidal sleeve of the trigeminal nerve. There is a wide medial connection to the premedullary cistern conveying the basilar artery; laterally it extends to the fundus of the internal auditory canal (cf. Fig. 7.2, 7.3 and 9.1 a). According to Lang (v. Lanz and Wachsmuth 1979), it forms a cushion ventral and cranial to the facial nerve in the region of the porus. Such topographical details are of decisive importance when the cranial nerves are detached from the tumor surface (Yaşargil 1969, 1984).

The *trigeminal nerve* emerges at the middle cerebellar peduncle. Several bundles of fibers are arranged for branches I through III in cranial-to-caudal order (Gudmundsson et al. 1971). The motor fibers emerge rostral or lateral to the pars major of the trigeminal nerve and subsequently run medial to its pars compacta and pars triangularis within Meckel's cavity.

The *abducent nerve* leaves the pontomedullary groove about 4 mm lateral to the midline and runs cranially and laterally toward its dural porus (see also Umansky et al. 1991).

The *facial nerve* proceeds from the lateral margin of the pontomedullary groove, immediately above the olivary body, and turns to the internal auditory canal together with the *vestibulocochlear nerve*, which originates further laterally and dorsally. Close to the brain stem it is impossible to differentiate between the portions of the eighth cranial nerve; exact subdivision can only be attained at the internal auditory canal. In the lateral suboccipital approach, one first encounters the superior and inferior vestibular nerves from a dorsal direction; they block one's view of the facial nerve, situated ventral to the superior vestibular nerve, and the cochlear nerve, which is located ventral to the inferior vestibular nerve. The *nervus intermedius* – the sensory root of the facial nerve – may change over from the vestibulochochlear nerve to the facial nerve along it's entire cisternal course (Rhoton 1968). In 33%, however, it leaves the brain stem between VII and VIII (Tschabitscher and Höcker 1991). The fibers of the intermedius nerve leave the facial nerve at the geniculate ganglion to form the greater petrosal nerve, and in the mastoid segment of the facial nerve as the chorda tympani (Fig. 7.4).

The fascicles of the *glossopharyngeal, vagus, and accessory nerves* proceed from the retro-olivary groove to the pars nervosa of the jugular foramen. The spinal root of the accessory nerve is conveyed by the foramen magnum and runs forward dorsal to the vertebral artery and dorsal to the topmost denticulate ligament (Fig. 8.5). The posterior inferior cerebellar artery usually passes through the root fibres of the caudal cranial nerves (Fig. 7.1).
The root fibres of the *hypoglossal nerve*, which emerge ventral to the olivary body, enter the canal of the hypoglossal nerve caudal to the jugular tubercle (Fig. 2.4). The distances between the seventh and eighth cranial nerves and the ridge of the petrous bone cranially and between the seventh and eighth cranial nerves and the vagus group caudally are almost equal, being 4 mm in each case.

The *inferior anterior cerebellar artery* branches off the basilar artery and in 80% of all cases examined runs laterally ventral to the abducent nerve; its relation to the seventh and eighth cranial nerves and the porus varies greatly (Sunderland 1945; Mazzoni 1969; Tschabitscher and Perneczky 1974; Perneczky 1979, 1980; Martin et al. 1980; Lang, 1981). According to Mazzoni (1969), the loop of the inferior anterior cerebellar artery is found in the internal auditory canal in one-third of cases, at the porus in one-third, and in the cerebellopontine angle in one-third (see. Figs. 7.3, 7.4, 9.1 b).

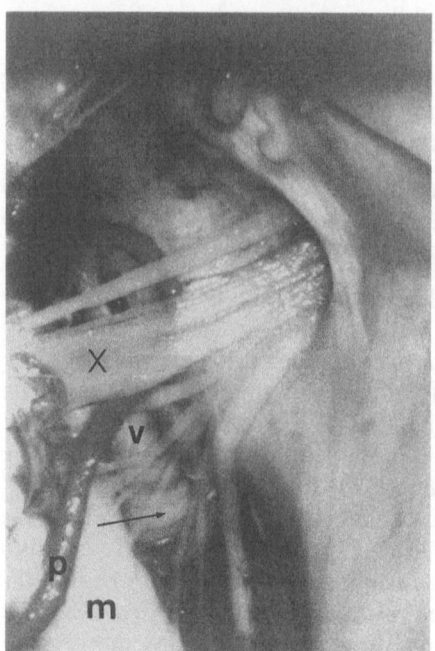

Fig. 7.1. Specimen in which the arterial system has been injected with dyed latex milk. The loop of the posterior inferior cerebellar artery (*p*) turns dorsally between the root fibers of the vagus nerve (*X*). The cisternal segments of – the ninth, tenth, and eleventh cranial nerves are visible as far the jugular foramen. The root fibers of the twelfth cranial nerve (*arrow*) are stretched around the vertebral artery (*v*). *m*, Medulla oblongata

The intradural segment of the *vertebral artery* up to the point of junction, is 25 mm long. The typical origin of the *inferior posterior cerebellar artery* is in the proximal third of the vertebral artery; less commonly it originates in the middle or distal third (Lanz and Wachsmuth 1979). Occasionally, the inferior posterior cerebellar artery may emerge from the basilar artery trunk. Its origin is very rarely extradural (Lang 1985a). From its origin the artery proceeds in a loop rising cranially; it is still located ventral to the caudal cranial nerves. In most cases, it then takes a dorsal turn between the root fibres of the accessory nerve (Fig. 7.1), or, less frequently, between the ninth and tenth or tenth and eleventh cranial nerves. The anterior and posterior inferior cerebellar arteries are in hemodynamic balance.

Fig. 7.4. The seventh and eighth cranial nerves have been exposed in a lateral suboccipital approach from the left. The vestibulocochlear nerve (*black arrow*) has been slightly displaced caudally with a dissector (*d*) for better exposure of the sensory root of the facial nerve – the nervus intermedius (*white double arrow*). The anterior inferior cerebellar artery (*a*) between the vestibulocochlear nerve and the sensory root of the facial nerve is unusually large. Note the small labyrinthine artery (*white arrows*). The cerebellomedullary cistern, where the ninth, tenth, and eleventh cranial nerves (*X*) are located, has not yet been opened. *Cb*, Cerebellum; *black double arrow*. facial nerve

Key Steps of the Approach

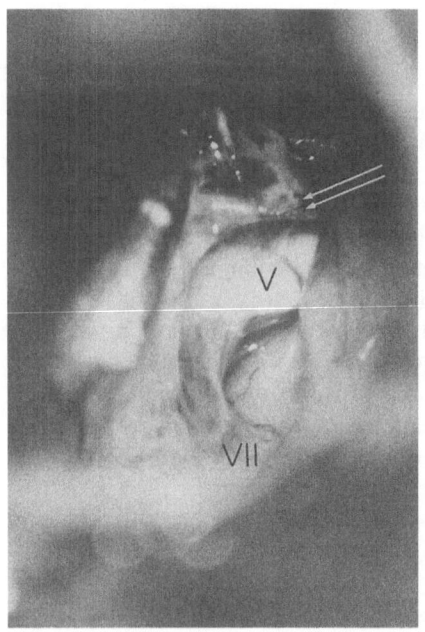

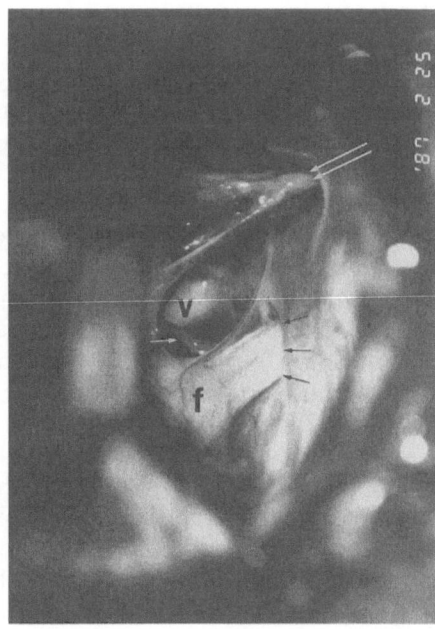

7.2

7.3

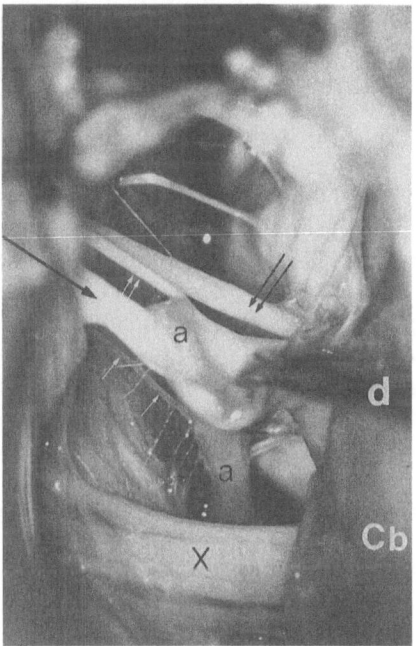

7.4

Fig. 7.2. Lateral suboccipital approach from the right. The arachnoid of the cistern at the cerebellopontine angle has not yet been incised, so the seventh and eighth cranial nerves are barely visible yet (*VIII*). The trigeminal nerve (*V*) and the petrosal vein (*white double arrow*) are more distinct already

Fig. 7.3. After the cistern has been opened, the facial and vestibulocochlear nerves can be exposed; they are covered by the flocculus (*f*). In the depths the trigeminal nerve (*V*), the petrosal vein (*white double arrow*), the anterior inferior cerebellar artery (*white arrow*) and the internal auditory canal (*black arrows*) are visible

Surgery

Technique

The patient is placed in a semisitting position with the legs propped up ("launching position") and the head turned approximately 30° toward the side on which the tumor is located. The skin incision is vertical and retromastoidal. The muscles are also incised vertically, i. e., parallel to the fibres of the trapezius muscle, and the occipital bone is exposed. Craniectomy is performed past the margins of the transverse and sigmoid sinuses and displaced caudally and medially. After the arachnoid has been cut, the cerebellum is carefully retracted medially, which is facilitated by aspiration of CSF when the cisterns are opened. If there is elevated intracranial pressure in the posterior fossa, it is advisable first to open the dura and arachnoid above the cerebellomedullary cistern. Thus, dural opening along the sinuses is less difficult and risky after CSF aspiration. We prefer this procedure to puncturing the lateral ventricle.

When the cerebellopontine angle is reached the tumor is exposed ventrally to the internal auditory meatus. If the tumor originates at the clivus, the seventh and eighth cranial nerves will be extended over its posterior aspect and must be identified prior to incision of the tumor capsule. If they are not perceptible because they are encased by the tumor or thinned, they can be identified at the internal auditory meatus and exposed from there, their arachnoidal sheaths being preserved if possible. The tumor capsule is then coagulated parallel to the course of the cranial nerves and incised. After purely endocapsular tumor reduction it will be easier to detach the seventh and eighth cranial nerves entirely from the tumor surface. After further tumor reduction, the caudal cranial nerves at the inferior aspect of the tumor, together with their arachnoidal sheaths, must be taken into consideration. While finding the positions of the seventh and eighth as well as the ninth, tenth, and eleventh cranial nerves is relatively easy, the same is by no means true of the abducent nerve: depending on tumor origin and direction of growth, this nerve may be situated at the upper ventral or lower ventral aspect of the tumor. It is, furthermore, endangered when the site of tumor origin, which is usually in the immediate area of its entry into the dura, is coagulated. Due to the closeness of their areas of entry into the dura, not only the ipsilateral but also the contralateral abducent nerve may be in contact with the tumor. Bilateral paralysis of the abducent nerve may also be the only neurological sign in patients with clival tumors (Kline and Glaser 1981). If the tumor passes through the tentorial notch, minor supratentorial tumor portions may be removed after the tentorium has been split. However, it was pointed out by Cushing (as quoted in Olivecrona 1967) that it was impossible to remove major supratentorial tumor portions by means of a lateral suboccipital approach. During this step, the fourth cranial nerve would be especially endangered because its point of entry into the dura of the anterior petroclinoid fold can not be made out from the posterior cranial fossa.

An important factor contributing to the success of surgery and surgical feasibility as such is the position of the tumor in relation to the basilar artery and its brain stem vessels. If the tumor surrounds the basilar artery or its branches and detachment is difficult, radical operation is unthinkable. This state of affairs has remained unchanged despite the advent of ultrasonic aspirators or laser – which are not maneuverable enough for this special area, anyway.

Case Report

A 45-year-old women complained chiefly of pain in the face which implicated the first two trigeminal branches on the right. In addition, the corneal reflex was reduced. Occasionally, complaints similar to a hemifacial spasm occurred; when turning her eyes to the extreme right, the patient was troubled by diploplia due to discreet paresis of the abducent nerve. CT revealed a giant prepontine epidermoid causing displacement and contorsion of the brain stem. The basilar artery was retracted far above the brain stem and displaced contralaterally. The space-occupying lesion stretched as far as the infundibulum and did not extend into the middle cranial fossa (Fig. 7.5a).

Surgery was performed via the lateral suboccipital approach on the right in the manner described above. The seventh and eighth cranial nerves were dis-

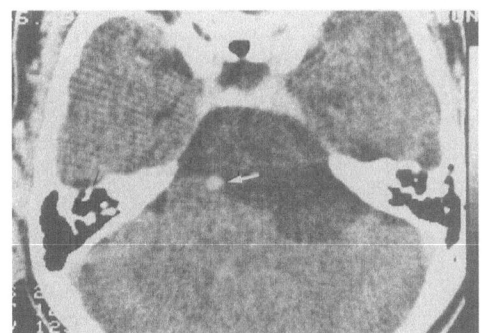

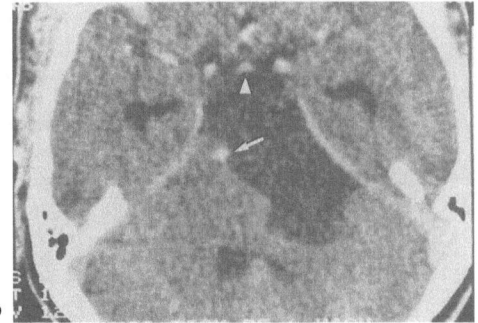

Fig. 7.5a, b. CT shows a marked prepontine space-occupying lesion caused by a intradural epidermoid. The brain stem and basilar artery (*arrows*) are contorted and displaced. The space-occupying lesion reaches far into the interpeduncular fossa and as far as the infundibulum (**b**, arrowhead)

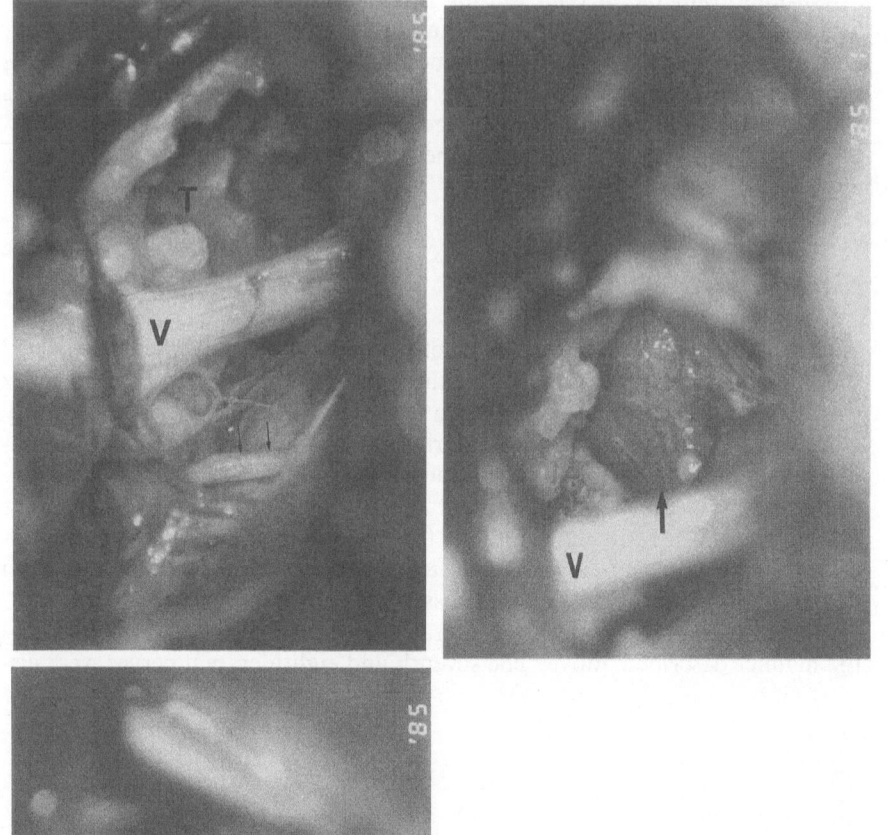

Fig. 7.6a–c. Lateral suboccipital approach from the right. **a** The trigeminal nerve (*V*) has been reached and the seventh and eighth cranial nerves (*arrows*) have been detached; a large portion of the epidermoid (*T*) is still left in the prepontine space. **b** Interpeduncular fossa: *V*, trigeminal nerve; *arrow*, infundibulum. **c** View of the tuber cinereum (*t*), contralateral optic tract (*arrow*), and contralateral posterior communicating artery (*double arrow*) at higher magnification

placed dorsally, the trigeminal nerve had been thinned and adjoined the tentorium. In the course of dissection, the right abducent nerve and later on the basilar artery and its branches were exposed. The space-occupying lesion extended into the interpeduncular fossa along the dorsum sellae, so a view not only of the contralateral oculomotor nerve but also of the infundibulum, the contralateral optic tract, and the contralateral posterior communicating artery was afforded (Fig. 7.6c). Apart from postoperative meningeal irritation, recovery was uneventful.

Frequently Encountered Lesions

The lateral suboccipital approach is primarily suitable for *meningiomas* that are situated mainly in the posterior cranial fossa and do not extend to the middle cranial fossa and/or do not involve the tentorium. The origin of such meningiomas is mostly the sphenopetroclival boundary; the approach is most appropriate for petroclival growths. Depending on site of origin and direction of growth of the tumor, the third, fourth, and fifth cranial nerves are displaced cranially; the trigeminal nerve is sometimes split up and infiltrated by the tumor, the abducent nerve is displaced medially, and the seventh and eighth cranial nerves are situated at the dorsal and caudal tumor circumference. As their mass is located lateral to the midline, such tumors displace the brain stem dorsally and contort it. This makes it possible to reach the midline and thus even the basilar artery and its branches by a lateral suboccipital approach. Purely medioclival meningiomas as described by Hakuba et al. in 1977 are very rare, or according to Yaşargil et al. (1980) do not occur at all.

When *chordomas* or *chondromas* of the clivus attain a certain size, they perforate the dura and extend into the prepontine space. In such cases one is confronted with tumors situated on both sides of the midline, which – due to the extension of the space-occupying lesion a long way laterally – can be extirpated neither by an anterior approach (Chaps. 1–4) nor by a lateral suboccipital approach. This desperate situation is reflected in the short survival times (Krayenbühl and Yaşargil 1975) and the futility of radiotherapy applied to the remaining tumor portions (Dahlin and MacCarty 1952; Arnold and Herrmann 1986). Still, the lateral suboccipital approach is a feasible way to partial tumor resection and brain stem decompression.

Epidermoids of the cerebellopontine angle are purely intradural lesions which often extend far into the prepontine space. They grow along the basal cisterns and settle in all their recesses. Their extirpation usually does not give rise to difficulties and they offer unusual insights into anatomy (see the case report in this chapter).

To a limited extent, the approach can also be used for *aneurysms* of the basilar trunk. However, without brain stem retraction it is impossible to gain sufficient access, so in these aneurysm cases we prefer the dorsolateral foramen magnum approach (see Chap. 8).

Advantages and Disadvantages of the Approach

The lateral suboccipital approach has the decisive advantage that it is a standard approach, i. e., from using it frequently, the surgeon knows both the limits of the approach itself and his or her own personal limits as to dissection. Another advantage is that from this approach the seventh and eighth cranial nerves are extended over the posterior tumor aspect and may be identified at an early stage. The same is true of the ninth, tenth, and eleventh cranial nerves, which are displaced caudally. In further dissection, this advantage may turn into a certain drawback, as tumor extirpation has to be performed again and again between the elongated bundles of cranial nerves, which have now been detached.

The option of extending the approach by performing posterior temporal craniotomy and turning it into a combined supra-infratentorial approach may be seen as a further advantage. It lends itself well to tumors involving the tentorium and/or tumors large portions of which extend into the middle cranial fossa (see Chap. 9). In tumors reaching the level of the foramen magnum, extension of the approach into a dorsolateral approach to the foramen magnum may be considered (see Chap. 8).

Insufficient access to even small portions of tumor in the middle cranial fossa must be regarded as a drawback of the approach. In a few cases, when the brain stem has been severely contorted, the structures at the midline may be reached. The lateral suboccipital approach alone, however, is not suited to extirpation of extradural lesions. Still, combined approaches offer a number of options (Fig. 7.7).

The discussion about the sitting position for exposure of the posterior cranial fossa and its advantages and drawbacks is as old as the technique itself (De Martel and Guillaume 1931; De Martel and Thurel 1936; Pool 1966; De Martel 1967 cited in Olivecrona 1967; p 6; Kurze 1979; Hitselberger and House 1980; Norell 1982). The positioning of the patient, however, does not affect the options offered by the lateral suboccipital approach.

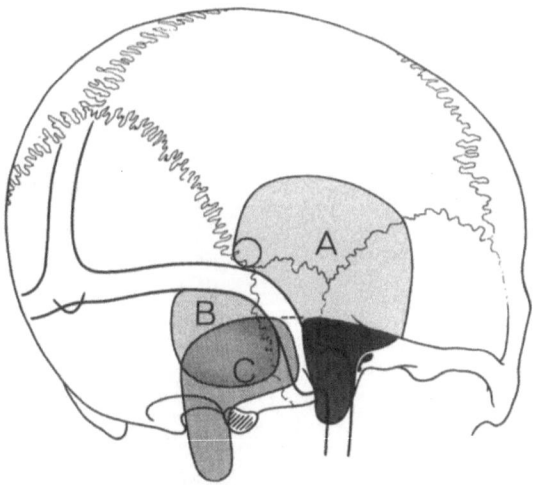

Fig. 7.7. Options for the combination of approaches: Classical subtemporal approach (*A*) with the posterior trepanation hole directly above the transversal sinus; lateral suboccipital approach (*B*) as far as to the sigmoid and transverse sinuses; dorsolateral approach to the foramen magnum (*C*); petrosal approach including resection of the petrous bone (area *shaded dark*)

Summary: Lateral Suboccipital Approach

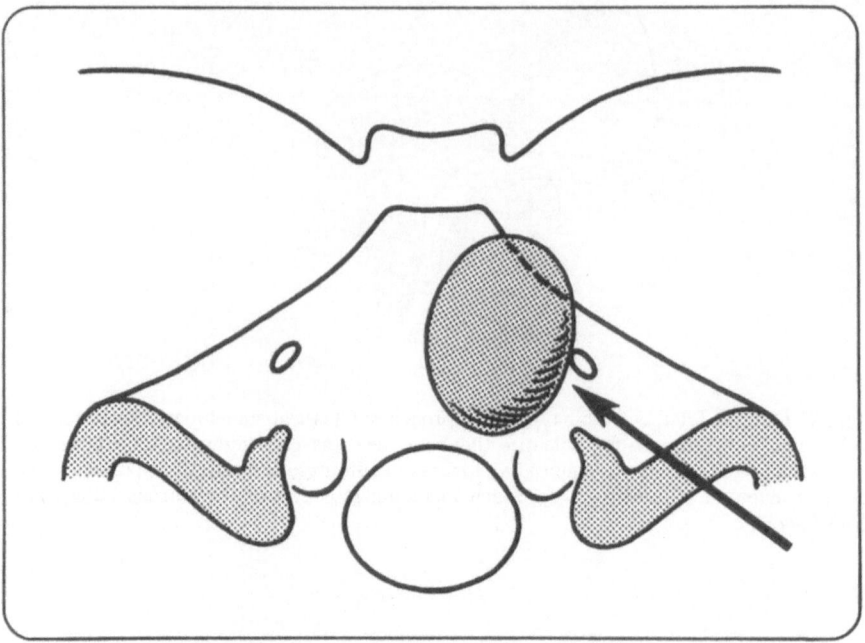

Indications: Petroclival tumors

Meningiomas – Trigeminal neurinomas – Epidermoids

Advantages: Standard approach
Seventh and eighth cranial nerves identified early
Possibility of combinations

Disadvantages: Only intradural tumors
Difficulty in reaching the middle cranial fossa
Cerebellar retraction

Limits: Midline
Tentorial notch

Chapter 8: Dorsolateral Approach to the Clivus and Foramen Magnum

Historical Survey

This modification of the lateral suboccipital standard approach was first mentioned in 1985 by Koos et al. (1985) and later on set out in detail by Perneczky (1986). The approach seems particularly suited for reaching lesions on the ventral rim of the foramen magnum or in the lower half of the clivus. This goes especially for so-called craniospinal meningiomas, which are usually situated at the anterior circumference of the foramen magnum. The approach also lends itself well to treating aneurysms of the vertebral artery, the vertebral junction, and the caudal basilar artery (Heros 1986).

Anatomy

Anatomy Relevant to Surgery

The *vertebral artery* passes forward in a cranial direction in the foramina transversaria of vertebrae C6–C3 and turns laterally and cranially to the transverse process of C2, which is situated far laterally (Fig. 8.3). After passing through the foramen transversarium of the atlas, the vertebral artery runs dorsally in a sagittal plane and at the arch of the atlas continues medially in its groove. At this point it passes immediately behind the atlanto-occipital joint and enters the dura of the craniocervical junction from dorsolaterally to ventromedially (Fig. 8.4). The groove for the vertebral artery (in rare instances, it develops in the form of a vertebral arterial canal) contains a more or less marked periarterial venous plexus. In its horizontal segment at the arch of the atlas, a muscular branch originates from the vertebral artery. If forms a connection to the occipital artery and fixes the position of the vertebral artery, so mobilization requires severing of the muscular branch (Fig. 8.10). Directly at area of entry through the dura, a *meningeal artery* branches off the vertebral artery; this must on no account be confused with an extradural origin of the posterior inferior cerebellar artery (Lang 1985). The topmost *denticulate ligament*, which runs through the foramen magnum, is attached immediately behind the area of dural entry of the vertebral artery and covers it from dorsally (Figs. 8.5, 8.6a). The accessory and hypoglossal nerves almost always run dorsal to the artery (Lang 1986b) (Fig. 8.1).

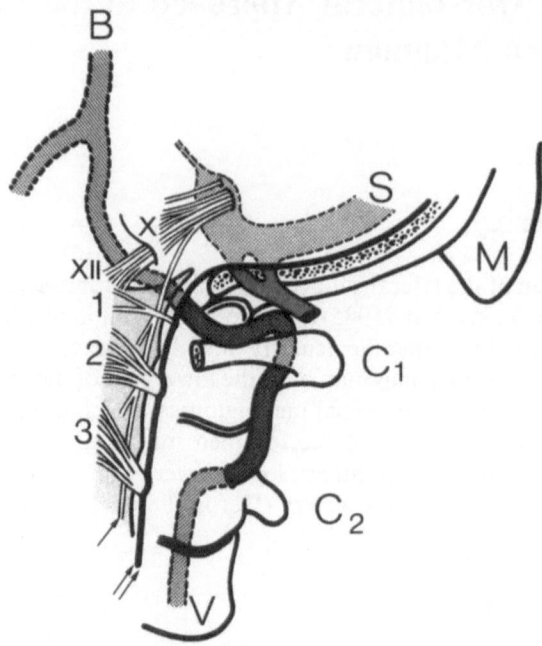

Fig. 8.1. Diagram of the vertebral artery (*V*) at the craniocervical junction. The segments between C1 and C2 as well as the atlas segment are *shaded dark*. Shortly after the dural entry area, the upper denticulate ligament (*very light shading*) covers the artery dorsally. The spinal root of the accessory nerve is situated dorsal to the denticulate ligament (*arrow*). The sigmoid sinus (*S*) and ninth, tenth, and eleventh cranial nerves (*X*) as well as a sturdy condyle emissary vein are shown. *XII*, Hypoglossal nerve; *1–3*, intradural segment of root fibers C1–C3; *M*, mastoid process; *B*, basilar artery; *double arrow*, dura

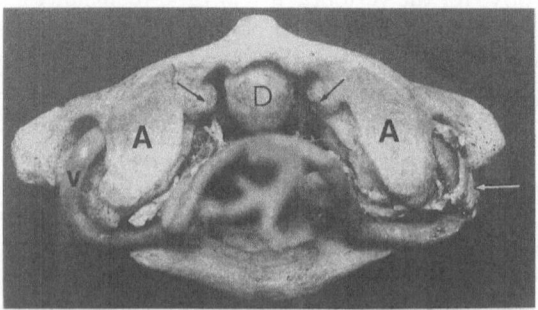

Fig. 8.2. Corrosion specimen of the upper cervical spine after injection of arterial and venous vessels (seen from above). The segment of the vertebral artery (*v*) at the arch of the atlas is visible, circling the atlanto-occipital joint (*A*) first on it's lateral then on it's dorsal and medial aspects. The periarterial venous plexus on the right (*white arrow*) has been better preserved. The tubercles which give attachment to the transverse ligament of the atlas are marked by *small arrows*. *D*, Odontoid process

The *anterior spinal arteries* normally originate a little short of the vertebral junction (5.8 mm, according to Lang 1986c) and join the contralateral artery after a short distance to form an impar vessel (see Fig. 1.6). The two vertebral arteries join to form the basilar artery at the level between the lower and middle third of the clivus which corresponds to the level of the pontomedullary sulcus.

Key Steps of the Approach

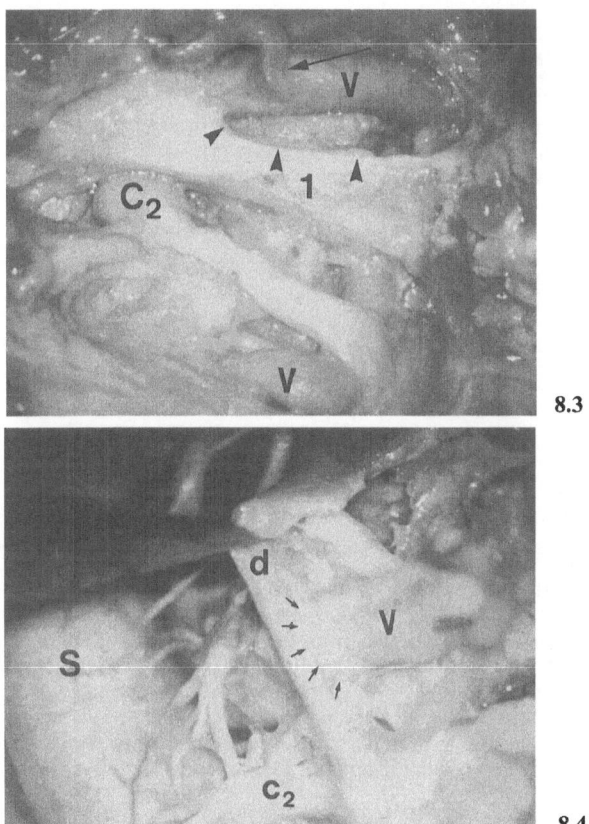

8.3

8.4

Fig. 8.3. The vertebral artery (*V*) can be shown at its sulcus (*arrowheads*) on the lamina of C1 (*1*) as well as at the level of C2 with the cervical nerve root C2 (*C₂*) dorsal to the artery. Note the large muscular branch of the vertebral artery at the C1 segment (*arrow*)

Fig. 8.4. The area of dural entry of the vertebral artery (*arrows*) with the extradural segment of the vertebral artery (*V*) *S*, Spinal cord; *C₂*, cervical nerve root C2. The dura (*d*) at the level of the foramen magnum is demonstrated with a forceps

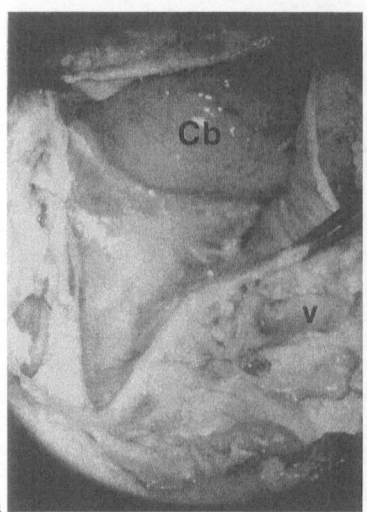
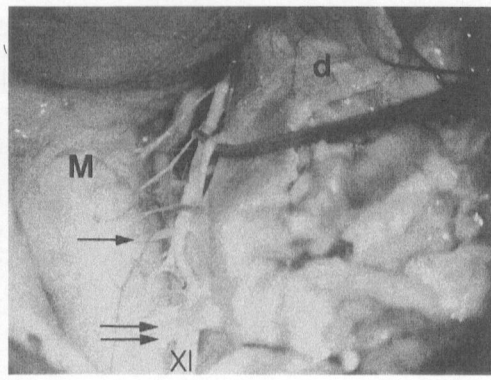

Fig. 8.5. a Dural incision at the level of the foramen magnum with the arachnoid over the medulla still intact. *Cb*, Cerebellum; *v*, vertebral artery. **b** The dura (*d*) is fixed by sutures and the arachnoid opened. This allows an approach ventral to the medulla (*M*). The topmost denticulate ligament is demonstrated with a hook. *Arrow*, C1 root entry zone; *double arrow*, C2 root entry zone. The accessory nerve (*XI*) passes dorsal to the denticulate ligament

Fig. 8.6. a View at the lower part of the approach with the accessory nerve (*XI*) ventral to the dorsal root C2 (*double arrow*) and dorsal to C1 (*black arrow*). *White arrow*, caudal loop of the posterior inferior cerebellar artery. **b** The cistern of the caudal crania nerves (*X*) has been opened and the choroid plexus is seen together with the vagal nerve. Anterior to the posterior inferior cerebellar artery, the rootlets of the hypoglossal nerve are visible (*arrow*). The *white arrow* shows the facial nerve group at the porus. **c** Between the vagal and the accessory nerve the pons (*p*) and the abducent nerve (*arrow*) can be seen. Note the characteristic relationship of the abducent nerve with the anterior inferior cerebellar artery (*double arrow*) dorsal to it

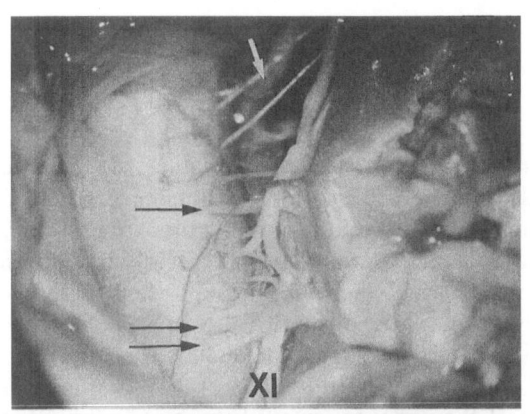

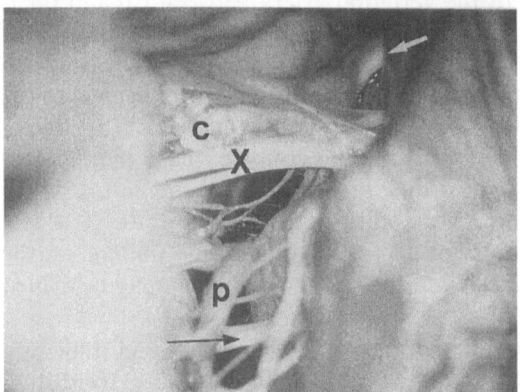

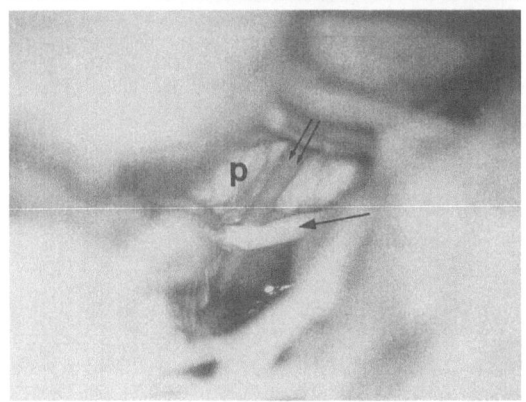

Surgery

Technique

The patient's position is similar to that used in the standard lateral suboccipital approach, seated, with slightly more head rotation and inclination. Head inclination for surgery of lesions ventrally to the medulla oblongata requires special care because the spinal cord and brain stem will already have been extended by the tumor. Skin incision and dissection are the same as in the retromastoidal approach, with additional exposure of the arch of the atlas. Craniectomy is performed to the rim of the sigmoid sinus; the foramen magnum has to be opened in its dorsolateral portion. The arch of the atlas is removed and the entire course of the vertebral artery between C0 and C1 is exposed extradurally from there. The key to this approach lies in drilling off the jugular tubercle extradurally with a diamond drill (Seeger 1980; Gilsbach and Seeger 1988), as it would obstruct the view of the structures ventral to the medulla oblongata (Fig. 8.7). Directly below the jugular tubercle the hypoglossal nerve runs through the base of the occipital condyle in the hypoglossal nerve canal; it is usually accompanied by large veins. These are connected to the jugular bulb as well as the prevertebral venous plexus. If there is still insufficient space to reach the anterior aspect of the medulla oblongata, the posterior part of the articular surface of the occipital condyle may be drilled off.

The dura is incised parallel to the sigmoid sinus, past the point of transition of the vertebral artery, and to the level of C1. At the point of dural transition,

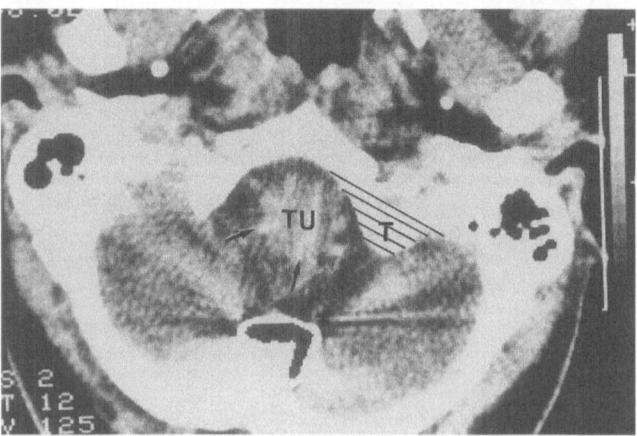

Fig. 8.7. Axial CT showing the tumor (*TU*) and a pronounced jugular tubercle (*T*) to be resected extradurally so that the ventral aspect of the foramen magnum can be attained without brain stem retraction

the vertebral artery usually gives off a dural branch. Dissection is continued along the vertebral artery, with the cerebellum being displaced upward rather than retracted from lateral to medial. Thus, the ventral aspect of the foramen magnum is reached without retraction of the brain stem.

Case Report

For years, this woman had suffered from neck pain and a certain clumsiness in her gait. During the year prior to surgery, clumsiness developed to virtual insecurity, which soon led to marked quadriparesis. Eventually, a lesion of the foramen magnum was diagnosed when rapidly progressive motor deficits appeared. Apart from motor weakness, the patient suffered from paresthesia in both arms. The large meningioma of the foramen magnum (Figs. 8.8, 8.9) was operated on the way described above. The tumor was polycyclic in appearance, which made dissection of the caudal cranial nerves difficult as they were encased by the tumor. The vertebral artery was displaced dorsally and embedded in the tumor at its area of dural entry (Fig. 8.11 a). After coagulation of the area of tumor attachment at the ventral margin of the foramen magnum, the blood supply was controlled and further extirpation was bloodless. It was

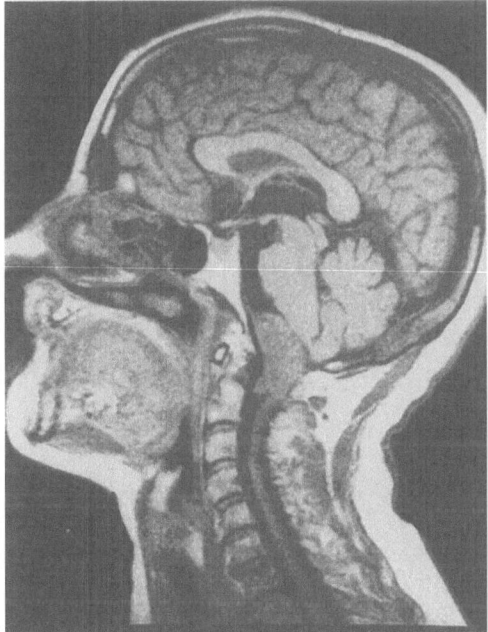
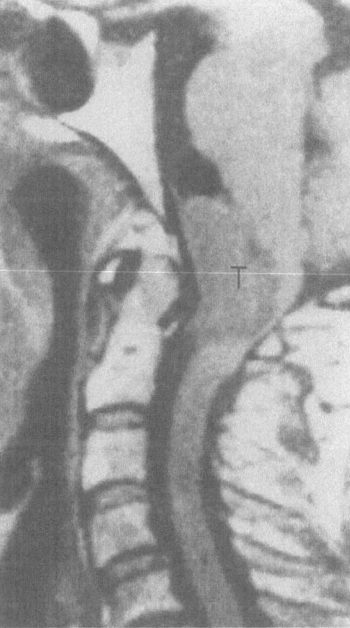

a   b

Fig. 8.8. a MRI of a typical ventral meningioma of the foramen magnum, the medulla oblongata being displaced and contorted. **b** Detail at magnification *T*, tumor

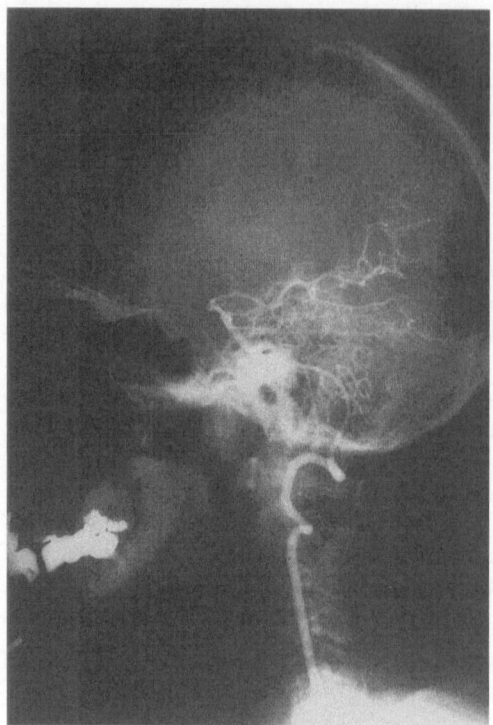

Fig. 8.9. Lateral view vertebral angiogram with the tumor, which is located ventrally, extending the vertebral and basilar arteries in an arch

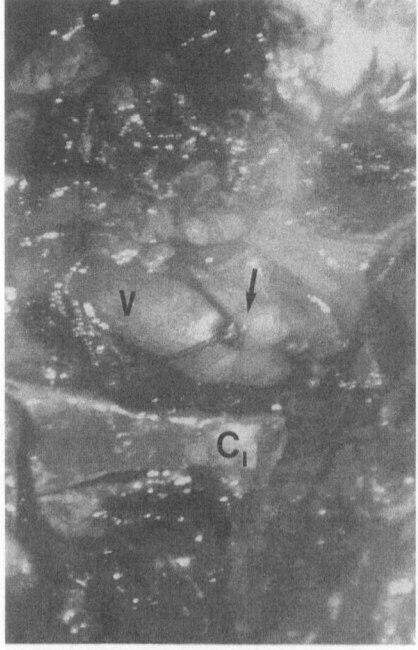

Fig. 8.10. Intraoperative view of the vertebral artery (v) at the arch of the atlas (C_1); *arrow*, ligation of a strong muscular artery

easy to detach the portion of tumor from the spinal cord. After complete
tumor removal, the junction of the two vertebral arteries and the origins of
both anterior spinal arteries were visible, as was the basilar artery as far as the
abducent nerve as the cranial (upper) limit.

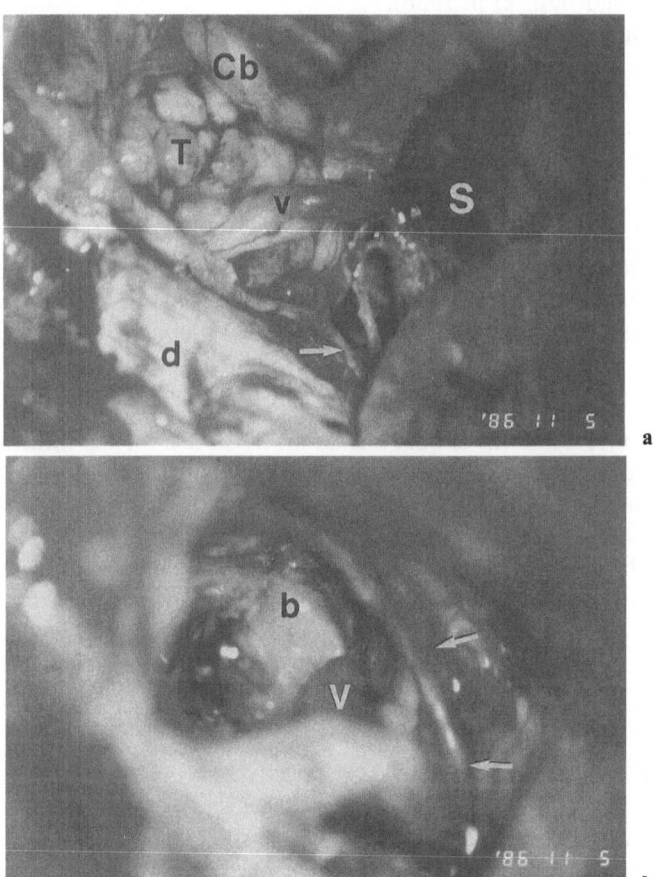

Fig. 8.11. a Dorsolateral approach to the foramen magnum from the left after opening of the
dura: the vertebral artery (*v*) is displaced in a dorsal direction by the tumor (*T*), which
surrounds it at the point of transition into the dura. *Cb*, Cerebellum; *arrow*, accessory nerve.
The dura of the caudal clivus has been fixed by staying sutures. **b** View of the resection cavity
after tumor extirpation: the vertebral artery (*v*) has been detached from the tumor; the basilar
artery (*b*) is visible in the depth and the posterior inferior cerebellar artery at the surface
(*arrows*)

Frequently Encountered Lesions

Most tumors of the craniocervical junction are *meningiomas* encroaching on the spinal canal as well as the posterior cranial fossa. Depending on the point of attachment to the dura, they are referred to as spinocranial (of spinal origin) or craniospinal (of cranial origin; Cushing and Eisenhardt 1938, on the basis of a suggestion by Bogorodinski in 1936).

Most of the foramen magnum meningiomas described by Stein et al. (1963) originate close to the area of dural entry of the vertebral artery. In many cases extirpation was impossible at that time, due to the extremely intimate relationship between tumor and artery. On the whole, only 2.5% of all meningiomas of the neuroaxis are meningiomas of the foramen magnum. In 85% of the cases examined their site of origin is the ventrolateral aspect of the foramen magnum (Stein et al. 1963). This agrees to our experience.

Neurinomas of the nerve roots of C1 and C2 may extend to the level of the foramen magnum: they made up one-third of the benign lesions of the foramen magnum described by Guidetti (1980, 1988). Zoltan (1974) reported that 25% of the tumors located at the foramen magnum were meningiomas, whereas 8% were neurinomas.

Aneurysms of the vertebral and the caudal basilar arteries are a further indication for the use of this approach. They make up approximately 2% of all intracranial aneurysms (Yaşargil 1984). Most frequently, they are located at the origin of the posterior inferior cerebellar artery and at the vertebral junction. In most cases, the dome of the aneurysm at the vertebral junction projects ventrally, so adhesion to the basal dura is frequently observed (Peerless and Drake 1982).

Advantages and Disadvantages of the Approach

The main advantage of this approach is that the anterior aspect of the foramen magnum can be attained from lateral and caudal without brain stem retraction. In space-occupying lesions of the ventral craniocervical junction, the medulla oblongata and/or spinal cord are dorsally displaced and contorted, so the rootlets of the hypoglossal and accessory nerves as well as of C1 and C2 are extended by the tumor. Transection of the upper denticulate ligament and, if that is still insufficient, of the rootlets of C1 and C2, reduces traction on the spinal cord and results in initial decompression. As dissection is carried out along the vertebral artery, the vessel is always under observation, which may be of special importance in meningiomas. We have already mentioned that the vertebral artery is nearly always in close contact with the tumor (Stein 1963; Cohen and Macrae 1962 personal observations). In one of the cases in which we perfomed surgery, the vertebral artery was encased by the tumor at its dural entry area; dissection of the tumor by the surgical technique described above, however, involved no major problems.

Derome and Guiot (1979) reported that, in an emergency, nondominant vertebral arteries may be ligated proximal to the origin of the posterior inferior cerebellar artery. After tumor removal, even the contralateral vertebral artery and the origin of the contralateral posterior inferior cerebellar artery are visible without the medulla oblongata having to be retracted. The dorsolateral foramen magnum approach is also very well suited for surgery on aneurysms of the vertebral artery, the vertebral junction, or the caudal basilar trunk as far as the level of the anterior inferior cerebellar artery (Heros 1986). When planning surgery on aneurysms of the vertebral junction, it is advisable to take an angiogram of both vertebral arteries. This is very important for the choice of the side of approach, as large or tortous vertebral arteries may completely cover small aneurysms; in such cases, access from the side of the non-dominant artery is recommended (Perneczky 1986). In large aneurysms of this location, however, the approach must be followed along the parent artery (to the aneurysm neck). Having had good experiences with using the dorsolateral approach in aneurysms of the vertebral junction and the caudal basilar artery, we see no great advantages in transoral or transcervical surgery. On the other hand, it is impossible to recognize and expose fenestrations of the basilar artery as the site of origin of aneurysms – as described by Matricali and Dulke (1981) and Black and Ansbacher (1984) – when using a dorsolateral foramen magnum approach. Adhesion of the aneurysm dome to the clival dura, which we have encountered in various cases, represents an additional reason for which we and others (Hoffmann and Wilson 1979; Peerless and Drake 1982) prefer to operate on such aneurysms by a posterior approach so as not to risk early rupture when opening the clival dura.

Summary: Dorsolateral Approach to the Clivus and Foramen Magnum

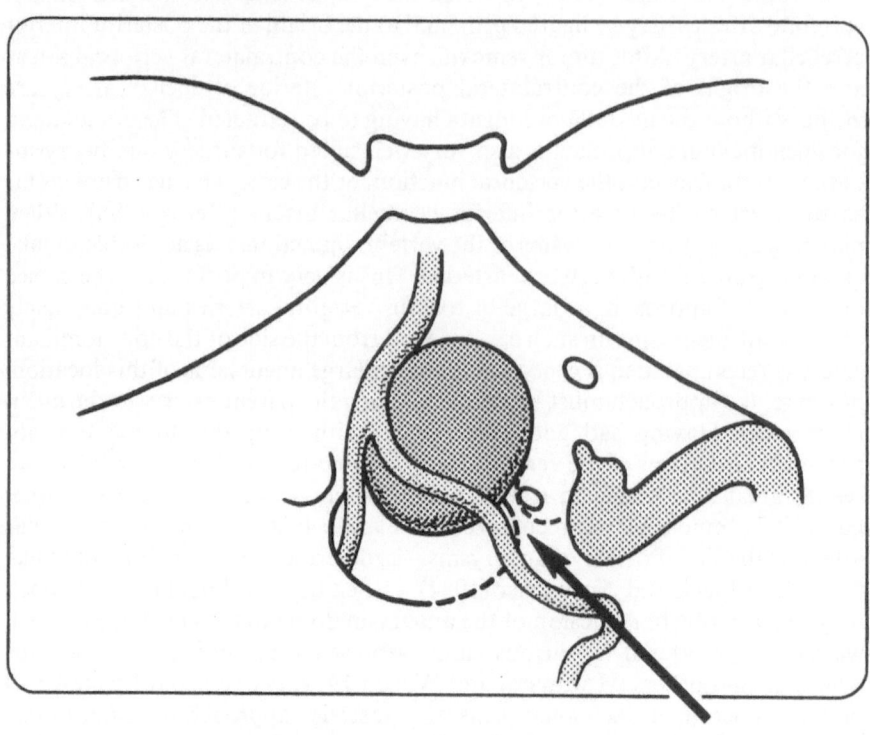

Indications:	Ventral and intradural tumors of the foramen magnum Aneurysms of the vertebral junction and proximal basilar artery

Meningiomas – Neurinomas – Aneurysms

Advantges:	No brain stem retraction Vertebral artery under control Dorsal decompression of the foramen magnum
Disadvantages:	Dissection between XI, XII, C1 and C2 Only intradural lesions
Limits:	Basilar artery Anterior inferior cerebellar artery Atlanto-occipital joint (?)

Chapter 9: Combined Subtemporal-Suboccipital Approaches

Historical Survey

In order to reduce the considerable retractor pressure in subtemporal-transtentorial approaches (Schisano and Tovi 1962) and still ensure adequate view of the posterior aspect of the petrous bone, the dorsum sellae, and midline structures, a posterior subtemporal approach may be combined with unilateral craniectomy of the posterior cranial fossa (Naffziger 1928; Fay 1930; Verbrugghen 1952; Bonnal et al. 1964; Luyendijk 1976; Pertuiset 1974; Samii 1981; Symon 1982; Malis 1984; Mayberg and Symon 1986; Al Mefty et al. 1988). In this way the surgeon is able to make use of the advantages of both approaches and to include a third advantage in the planning, that of gaining additional space in the region of the petrous bone. In our opinion, this combination is the most suited approach to petroclivotentorial lesions (Samii et al. 1989). Problems relating to temporal lobe as well as the cerebellum, which we wanted to avoid, made us choose this combined approach, the greatest advantage of which is that the temporal retractor hardly has to be used. Increasing experience with this approach and growing motivation brought by good results Samii et al. (1989) have prompted us to describe this variant of the combined supra-infratentorial approach separately and to compare it to the conventional approach as a "Variant B".

On the basis of the translabyrinthine-transtentorial approach of Morrison and King (1973) and the approaches described by Hakuba (Hakuba 1986; Hakuba and Nishimura 1981; Hakuba et al. 1977), we adapted these approaches to microsurgical requirements, modified them, and drew up a detailed report on the resulting approach. As the approach is basically characterized by opening of the dura of the posterior fossa presigmoidally and the direction of the approach follows the petrous ridge, we have called this the "presigmoidal petrous ridge approach".

Of course, it was not so much the tentorium which hampered the development of a combined subtemporal-suboccipital approach as the problems encountered in the ligation of the transverse sinus. Growing experience with the combined subtemporal-suboccipital approach has shown that ligation of the transverse sinus, once so dreaded, by no means inevitably entails disastrous consequences (Symon 1982; Mayberg and Symon 1986).

Anatomy

Anatomy Relevant to Surgery

For topographical correlations and special features of the tentorium, please also refer to the chapter on temporal approaches (Chap. 6).

Bridging veins from the temporal lobe most frequently insert at the junction of the transverse and sigmoid sinuses, less frequently in the horizontal segment of the transverse sinus, and rarely in the dura of the middle cranial fossa (Lang 1981). There are also descriptions of a short common segment in the tentorium before the exit into the transverse sinus (Oka et al. 1985).

Vein of Labbé: By definition, the vein of Labbé is an anastomotic vein between the superficial middle cerebral artery and the inferior temporal veins. For the neurosurgeon, the vein of Labbé is a large temporal vein draining into the transverse sinus (notwithstanding its anastomoses). This produces not only semantic problems, but also considerable "temporal lobe problems". All the veins not anastomosing should be called inferior temporal veins. The inferior anastomotic vein passes forward along the convexity of the temporal lobe and drains into the transverse sinus. Its exit may be situated close to the transitional zone between the transverse and sigmoid sinuses or further dorsally in the transverse sinus (Fig. 9.5). Reportedly, inferior anastomotic veins develop more frequently in the dominant hemisphere (Di Chiro 1962), and are found in 42%–55% of cases considered (Krayenbühl and Yaşargil 1979; Di Chiro 1962). In the tentorium, so-called tentorial sinuses have developed; they frequently drain cerebellar as well as cerebral veins to the transverse sinus. Drainage patterns at the torcular vary. Equal bilateral drainage as described in the textbooks is found in 57% of cases. In 25% of cases the superior sagittal sinus drains to one side and the tentorial sinus contralaterally. In 9% of cases a bifurcation of the superior sagittal sinus before the torcular is found (according to Bisaria 1985). In 60%–70% of cases, the transverse sinus on the right develops more markedly (v. Lanz and Wachsmuth 1979; Oka et al. 1985), which is explained by direct venous drainage to the heart.

Key Steps of the Approach: Variant A

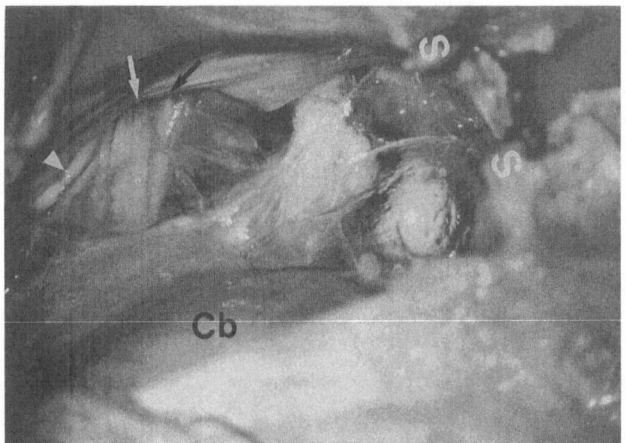

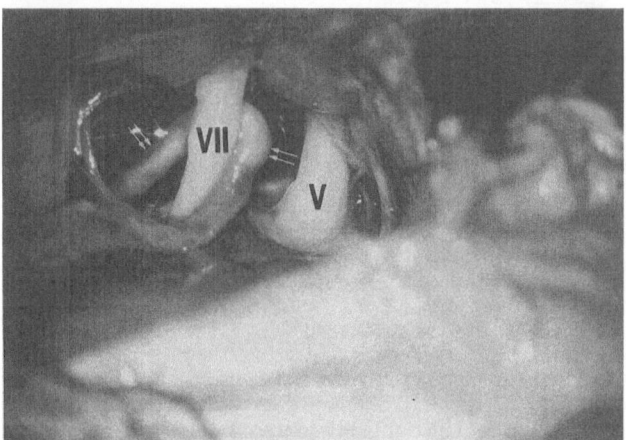

Fig. 9.1 a, b. Approach from the left in the lateral position. After craniectomy of the posterior cranial fossa and osteoplastic subtemporal craniotomy, the transverse sinus is ligated at the junction to the sigmoid sinus and transected (*S–S*). The tentorium is also transected along the ridge of the petrous bone. The cerebellum (*Cb*) falls away by gravity and CSF aspiration; thus, the cisterns of the cerebellopontine angle and the ninth, tenth, and eleventh cranial nerves are exposed without retraction. In **a** the arachnoid has not yet been incised but the cranial nerves are already visible. After the arachnoid has been opened (**b**), the seventh and eighth cranial nerves (*VII*) and a very large anterior inferior cerebellar artery (*white double arrows*) are exposed, as are the trigeminal (*V*), glossopharyngeal (*black arrow*), vagus (*white arrow*), and accessory (*white arrowhead*) nerves

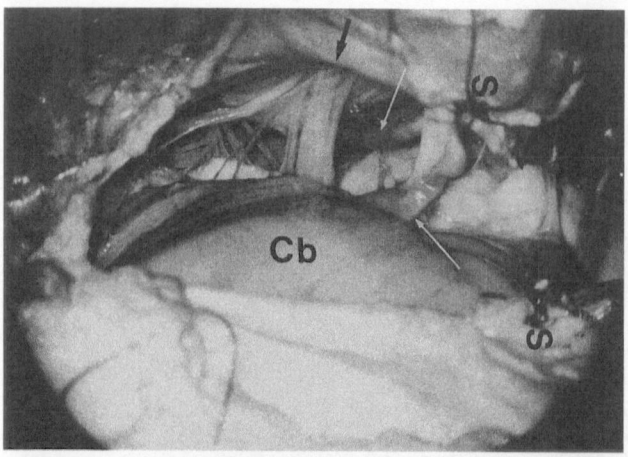

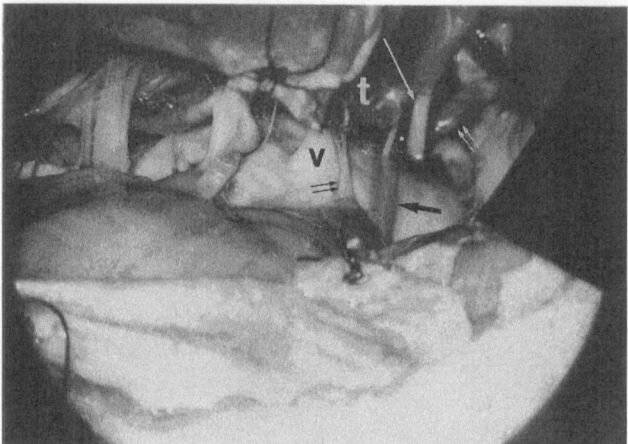

Fig. 9.2. a The caudal cranial nerves, among them the ninth, tenth, and eleventh cranial nerves (*black arrow*) and the hypoglossal nerve originating in several fiber bundles (*black double arrow*) are exposed. *White arrows,* anterior inferior cerebellar artery; *S-S,* ligated and transected transverse sinus; *Cb,* cerebellum. **b** Region of the tentorial notch, showing the trigeminal nerve (*V*) delicate petrosal vein (*black double arrow*), superior cerebellar artery and trochlear nerve (*black arrow*), which enters the dura of the anterior petroclinoid fold. *White arrow,* oculomotor nerve; *t,* rim of the tentorial notch; *white double arrow,* large posterior communicating artery

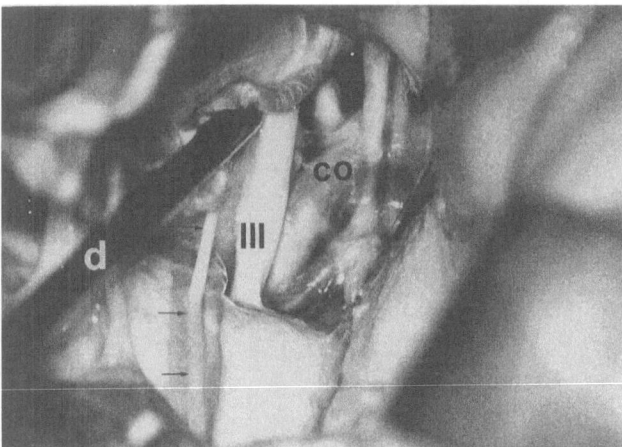

Fig. 9.3. View of the supra- and parasellar space. The anterior petroclinoid fold is retracted with a dissector (*d*); thus, the entry of the oculomotor nerve (*III*) into the lateral wall of the cavernous sinus is exposed. *Small black arrows*, Trochlear nerve parallel to superior cerebellar artery; *co*, posterior communicating artery

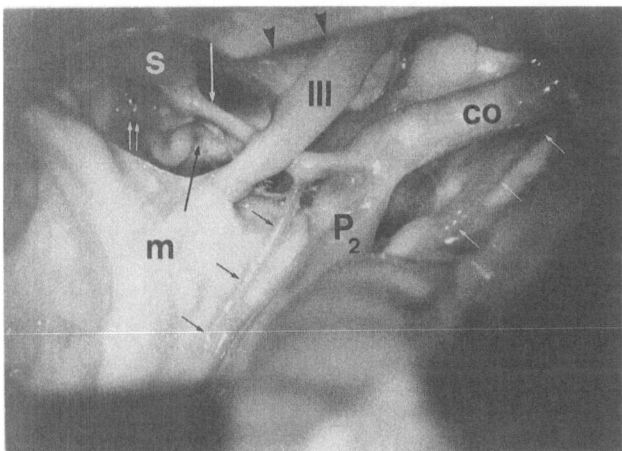

Fig. 9.4. Subtemporal view of the prepontine space. The large posterior communicating artery continues as a posterior cerebral artery of the embryonic type (*co*). The precommunicating segment of the posterior cerebral artery (P1 segment) is hypoplastic (white arrow). *S,* Superior cerebellar artery. The tortuous contralateral P1 segment of the posterior cerebral artery, which is also hypoplastic, drains into the contralateral posterior communicating artery (the *black arrow* marks the entry area). A large trunk exits the basilar bifurcation; it is in turn the site of origin of the posterior inferior diencephalic branches (*white double arrow*). *Small black arrows,* medial posterior choroidal artery branching off the posterior cerebral artery; *III,* oculomotor nerve; *black arrowheads,* anterior petroclinoid fold; *m,* lateral view of the midbrain; P_2, postcommunicating segment of the posterior cerebral artery, *small white arrows,* anterior choroidal artery

Key Steps of the Approach: Variant B

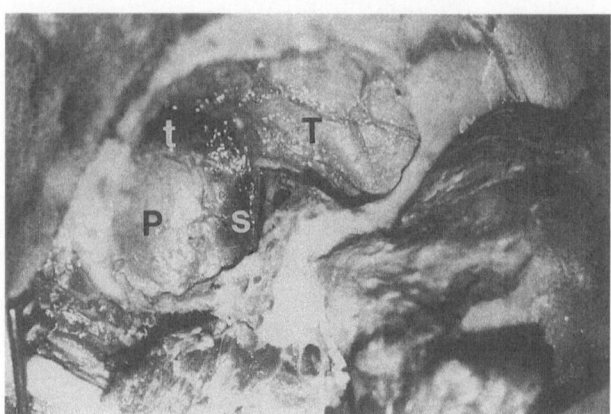

Fig. 9.5. Supra-infratentorial craniotomy on the right side as described for variant B. The specimen is positioned as for the semisitting position. The lateral parts of the petrous bone have already been removed. *T,* Temporal lobe; *P,* posterior fossa; *t,* Transverse and *s,* sigmoid sinus

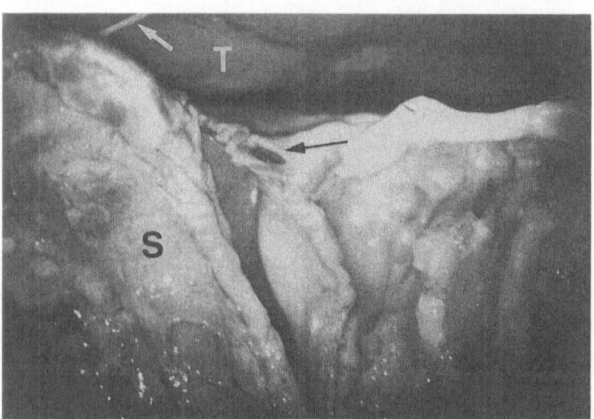

Fig. 9.6. The temporobasal dura has been opened and the temporal lobe elevated to demonstrate the vein of Labbé (*white arrow*). The dura of the posterior fossa has been opened presigmoidally with the superior petrosal sinus divided (*black arrow*). *T,* Temporal lobe; *s,* sigmoid sinus

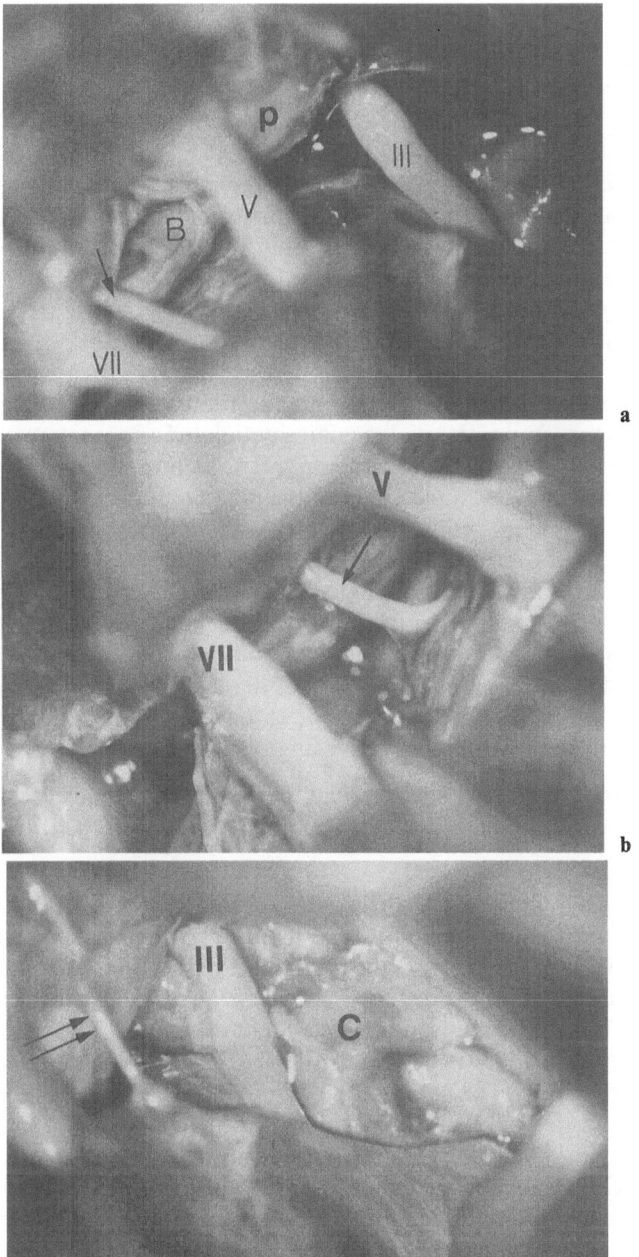

Fig. 9.7 a–c. View into the supratentorial (**a**) and infratentorial (**b, c**) space. *III*, oculomotor nerve; *double arrow*, trochlear nerve; *V*, trigeminal nerve; *arrow*, abducent nerve; *VII*, the facial nerve group, *C*, carotid artery; *B*, basilar artery; *p*, pons

Surgery

Technique for Variant A

The patient is for preference placed in the "park bench" position, with the head turned an additional 30° facing the floor. This ensures a equally good view of the posterior and middle cranial fossae. The approach to the clivus is centered along the petrous ridge. The skin incision is an extension of a retro-mastoid incision, describing a high temporal arc, approximately following the attachment of the temporal muscle. We usually start out with temporal trepa-nation, which must reach the floor of the middle cranial fossa. If the tumor has attained the internal carotid artery, the cavernous sinus, or the optic nerve, craniotomy may be extended pterionally beyond the ridge of the sphenoid bone. The bone flap at the temporal muscle remains petiolated and is retracted frontally and basally so as not to obstruct subtemporal dissection. Craniotomy is continued osteoclastically. The craniectomy of the posterior cranial fossa is carried out across the transverse sinus and as far as the lateral wall of the sigmoid sinus. The parts of the petrous bone adjoining the sigmoid sinus are drilled off in order to improve the visual angle. This is of the utmost impor-tance when structures above the level of the dorsum sellae have to be reached. In most cases, the most caudal portions of the squamous part of the temporal bone have to be drilled off to minimize retraction of the temporal lobe. When the dura is incised, the problems posed by veins have to be tackled; these need to have been already examined in preoperative angiography:

1. Patency of the contralateral transverse and sigmoid sinuses and jugular vein
2. Identification of the point where the vein of Labbé drains into the trans-verse sinus

The supratentorial dura is opened parallel to the transverse sinus; first the points of drainage of the vein of Labbe or large temporal bridging veins are identified. The dura of the posterior cranial fossa is also incised parallel to the transverse sinus and then following the sigmoid sinus. Thus, the transverse sinus is exposed from both sides and can be ligated and transected. The point of transection is usually close to the zone of transition to the sigmoid sinus, and is at any rate distal to the point of drainage of the vein of Labbé. The tentorium is transected from lateral to medial along the ridge of the petrous bone. Thus, the temporal lobe can be retracted together with the transverse sinus and the veins draining into it (Fig. 9.8 a). When the positions of the temporal retractor are changed, special care must be taken not to block the drainage of the vein of Labbé for long.

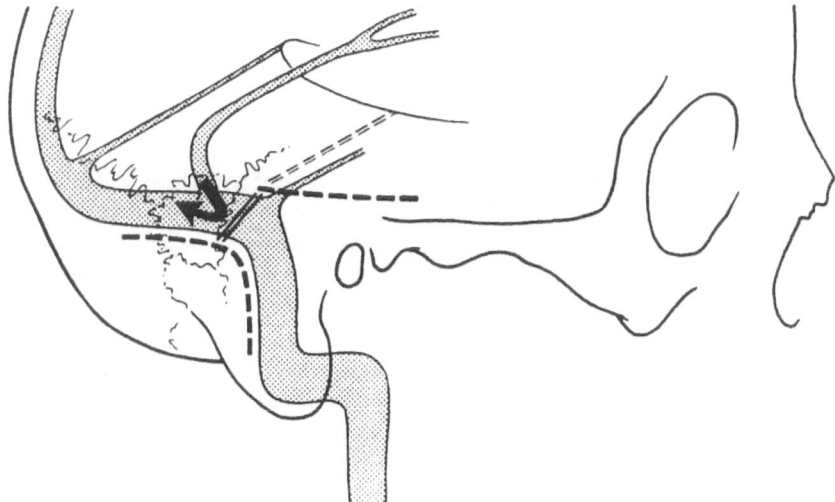

Fig. 9.8. a Dural incision with transsection of the sigmoid sinus (Variant A). The sigmoid sinus should be transected at the junction with the transverse sinus to allow the basal temporal veins to drain into the transverse sinus. Thus, the temporal lobe may be retracted together with the transected tentorium and the transverse sinus which drains the temporobasal veins towards the torcular (*curved arrow*). *Broken lines*, Supratentorial and infratentorial dura incisions; *Broken double line*, incision of the tentorium parallel to the petrous ridge and the superior petrosal sinus

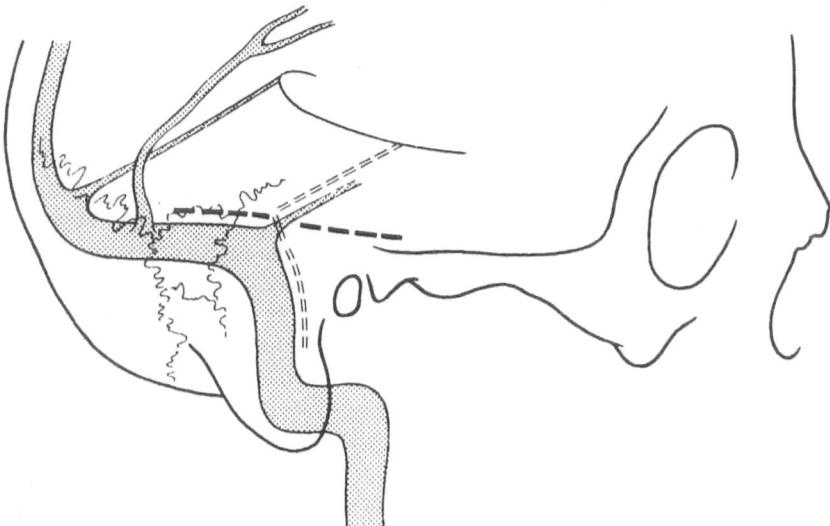

b Dura incision with preservation of the sigmoid sinus (variant B). The supratentoral duraincision is carried out along the temporobasal dura and continued along the transverse sinus until the vain of Labbé is identified (*broken line*). After the mastoid is drilled out, the anterior aspect of the sigmoid sinus and the dura of the posterior surface of the petrous bone can be seen; the dura of the posterior fossa is opened presigmoidally. After transection of the superior petrosal sinus, the tentorium is cut along the transverse sinus as well as along the petrous ridge from lateral to medial (*broken double line*)

Technique for Variant B:
The Presigmoidal Petrous Ridge Approach
(Modification according to Samii)

In variant B of the combined supra-infratentorial approach, we usually perform surgery with the patient in sitting position with the head turned by an
additional 30° as described above. The skin incision is an arced extension of
the retromastoidal incision leading around the ear in a semicircle to the zygomatic arch. The skin flap is dissected to as far as the external auditory meatus,
which exposes the posterior temporal squama, the occipital bone, and the
mastoid. Craniotomy is performed in such a way that the transition zone
between the transverse and the sigmoid sinus forms the center of the approach.
The temporobasal drillhole is placed above the transverse sinus and allows one
to perform a craniotomy no more than 1–1.5 cm above the sinus. This allows
sufficient exposure. After lateral suboccipital craniectomy (or craniotomy),
the sigmoid sinus has to be exposed in its entirety. The jugular bulb forms the
caudal limit of this approach. Once the sigmoid sinus is exposed, at least the
posterior portion of the mastoid has to be removed as well. The extent of
further resection depends on audition and the location of the process.

In cases involving complete loss of hearing, resection may extend as far as
the internal auditory meatus without regard to the labyrinth. The facial nerve
is left in its bony canal and forms the limit of petrous bone resection.

Where hearing is intact or the chance of improvement after tumor removal
exists, the labyrinth should at all costs be preserved. This also means that large
mediotemporobasal lesions are more difficult to reach, particularly, if they are
additionally located at the apex of the petrous bone and the clivus extradurally
(see also Figs. 5.1–5.3). First the dura is opened temporobasally (Figs. 9.8 b,
9.10 b) in a dorsal direction, so the drainage of the vein of Labbé is well visible.
The dura of the posterior fossa is opened presigmoidally and close to the sinus.
After the superior petrosal sinus has been divided, the tentorium is transected
along the petrous ridge as far as the tentorial notch or posterior rim of the
tumor. In addition to that, the temporobasal dura has to be sutured so that the
space gained by resection of the petrous bone can be utilized to the maximum.

Tacking sutures, and retractors draw the sigmoid sinus dorsally, the lateral
temporal dura cranially, and the dura of the petrous bone caudally. The lesion
can now be exposed by a minimum of cerebellar and temporal retraction and
thus extirpated (see Case Report below).

Before the tentorial notch is reached, the trochlear nerve requires special
attention. It is usually situated behind the dural margin and enters the dura at
the posterior petroclinoid fold (see Fig. 9.3). This area of entry through the
dura is where the nerve is most endangered. The trochlear nerve is woven into
the arachnoid in the region of the ambient cistern and may be mobilized along
with it when the temporal lobe is elevated. As the tumor has usually reached
the tentorial notch, the trochlear nerve is extended laterally and will cross the
field of operation during the entire duration of surgery. It is therefore in many

cases impossible to preserve it. Particular care should also be taken with regard to the oculomotor nerve, which is usually found at the lateral superior aspect of the tumor. It can be detached from the tumor left in its arachnoidal sheath. Minimized retraction on the cerebellum, which falls back, exposes the cerebellopontine angle and the facial nerve group, which is located lateral to the tumor. If the seventh and eighth cranial nerves are not visible on the tumor surface, they should be looked for at the internal auditory canal or their origin at the brain stem (depending on the stage of dissection) and dissected from there.

In meningiomas extending from the clivus, the trigeminal nerve is sometimes displaced laterally, but often split up by the tumor. In the latter case, functional preservation is impossible.

Displacement of the abducent nerve can be predicted in rare instances only. Depending on the site of tumor origin and direction of growth, the abducent nerve may be medially displaced or located at the anterior or the inferior aspect of the tumor. The brain stem is reached by piecemeal tumor resection. The angiogram will reflect what type of displacement the basilar artery and its branches have undergone. It is, however, impossible to predict whether they have only been displaced or are also encased by the tumor. Variations in diameter detected in the angiogram of the basilar artery or its major branches in the presence of adequate filling are a lead that the vessels are encased by the tumor. MRI will give better information on that all important detail. If the tumor causes little contortion of the brain stem and displaces it only dorsally, the contralateral internal auditory canal may be visible after tumor extirpation if circumstances are favorable (Symon 1982).

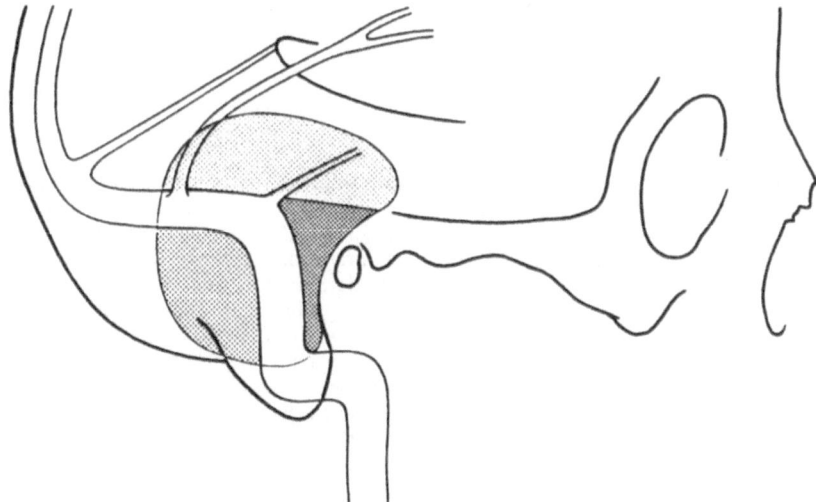

Fig. 9.9. The approach is centered on the petrous ridge, with supratentorial craniotomy, (*light shading*), infratentorial craniotomy (*medium shading*), and removal of petrosal bone (dark shading)

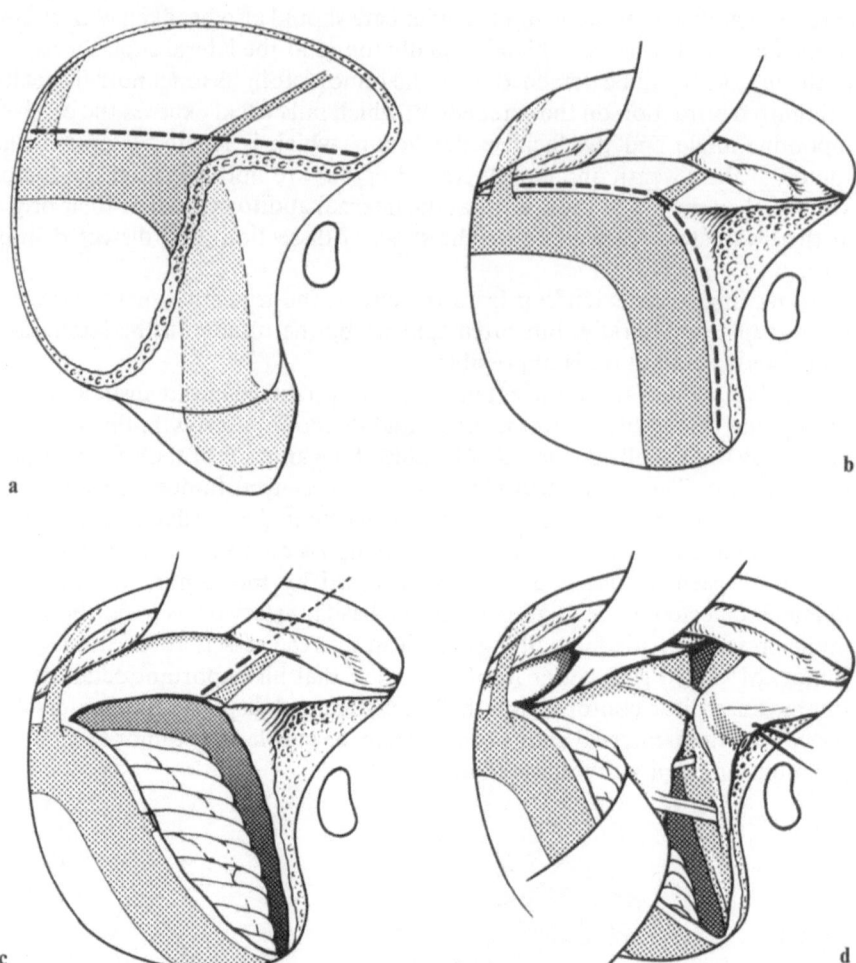

Fig. 9.10 a–d. The combined presigmoidal petrous ridge approach in detail. **a** The supratentorial craniotomy is carried out over the transverse sinus, followed by posterior fossa craniectomy. **b** The mastoid over the sigmoid sinus and, depending on hearing function, parts of the petrous bone have been removed (see text). The dura of the basal temporal lobe is opened, as shown in **a,** with the temporal retractor slightly elevating the temporal lobe. This maneuver allows identification of the vein of Labbé and opening of the dura along the anterior margin of the sigmoid and transverse sinus (*broken line*). **c** After transection of the superior petrosal sinus, the tentorium is cut parallel to the petrous ridge from lateral to medial (*broken line*). **d** After tentorium transsection, the dura over the petrous bone can be fixed by sutures, giving an excellent view into both the posterior fossa and the middle fossa, showing the third to eighth cranial nerves (see Fig. 9.7). Note that the temporal retractor has been placed beneath the tentorium to protect the temporal lobe

Case Report: Variant A

This 30-year-old woman (Figs. 9.11–9.13) had for years been suffering from deteriorating hearing on the left, occasional dysphagia, and hoarseness. In addition, there was palsy of the left accessory and hypoglossal nerves as well as progressive visual deterioration (over approximately 1 year). CT revealed a giant, enourmously vascularized tumor reaching from the sella to the level of C2. The vascular pattern was typical of a glomus jugulare tumor. Preoperative embolization was carried out a few days prior to surgery. Due to pronounced

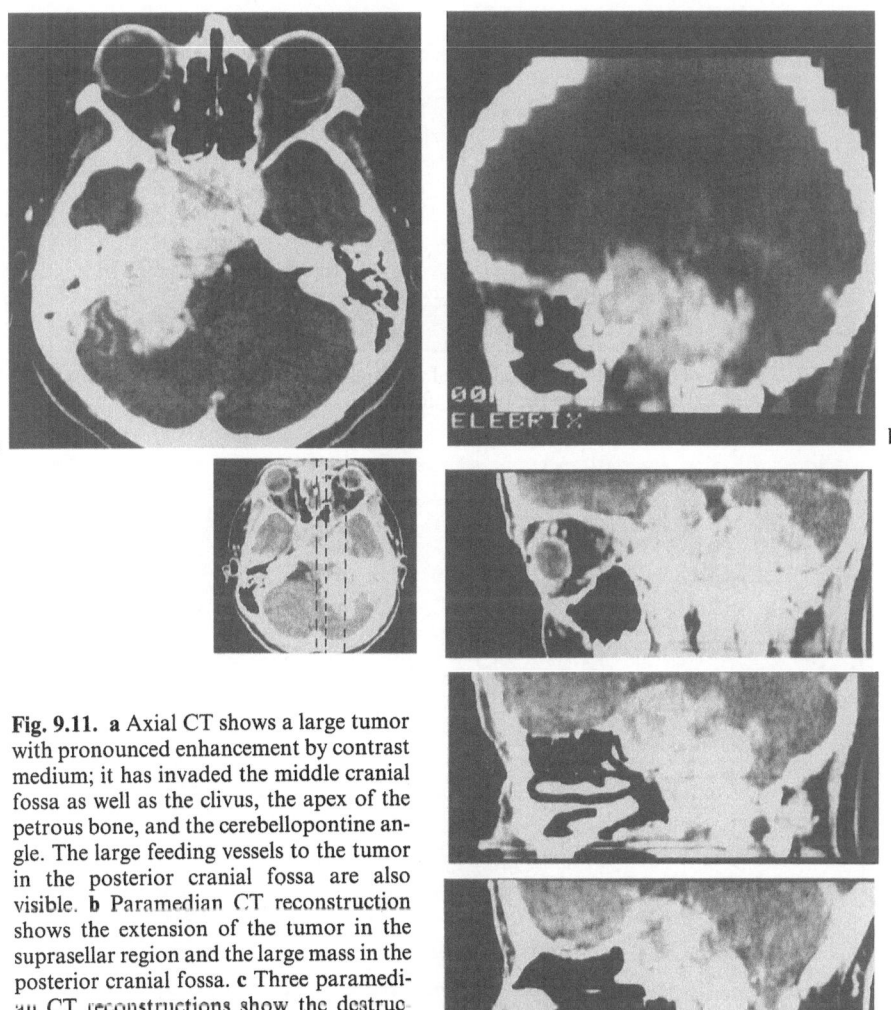

Fig. 9.11. a Axial CT shows a large tumor with pronounced enhancement by contrast medium; it has invaded the middle cranial fossa as well as the clivus, the apex of the petrous bone, and the cerebellopontine angle. The large feeding vessels to the tumor in the posterior cranial fossa are also visible. **b** Paramedian CT reconstruction shows the extension of the tumor in the suprasellar region and the large mass in the posterior cranial fossa. **c** Three paramedian CT reconstructions show the destruction of the petrous bone and the bone destruction at the clivus

posterior fossa involvement, the patient was placed in a semisitting position. The skin was incised in a large curve from the temple to the level of C2. Unlike what was described above, in this case the size of the tumor necessitated mastoidectomy together with labyrinthectomy. The facial nerve, however, remained in its osseous canal. The sigmoid and transverse sinuses had been almost completely occluded by the tumor, the transection was not expected so in all probability to have any hemodynamic sequelae. After the tentorium had been split, it was retracted, together with the temporal lobe.

Tumor removal was begun in the cerebellopontine angle on the left. Dissection was then continued toward the foramen magnum after the facial and vestibulocochlear nerves had been exposed and anatomically preserved. After

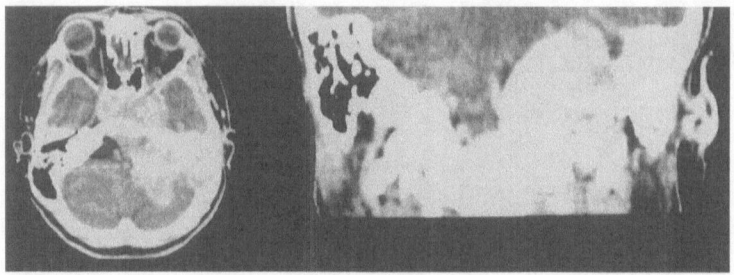

Fig. 9.12. Frontal reconstruction at the level of the jugular foramen; in the posterior cranial fossa, the tumor extends as far as the jugular foramen

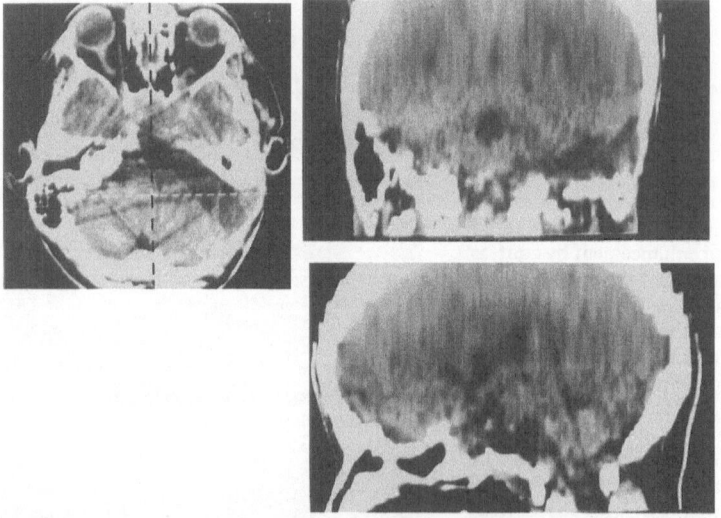

Fig. 9.13. Postoperative CT in median and coronal reconstruction: median reconstruction shows the extent of destruction at the osseous clivus

the caudal cranial nerves had been exposed and anatomically preserved, the tumor, which extended to the level of C2, was removed in a piecemeal manner. During tumor dissection, the vertebral and basilar arteries and their branches were attained. The oculomotor nerve was exposed between the superior cerebellar and posterior cerebral arteries and followed as far the cavernous sinus in its further course. As the cavernous sinus was entirely invaded by the tumor, the lateral wall of the sinus had to be resected, the trochlear nerve having to be sacrificed. It was possible to anatomically preserve the abducent and trigeminal nerves. The tumor extended as far as the intracavernous segment of the internal carotid artery. In the final stage, the optic nerves and the contralateral oculomotor nerve were dissected from the tumor in the suprasellar region. Postoperative transient dysphagia, which required a prolonged period of recovery, complete deafness, and peripheral facial paresis occurred despite the anatomical preservation of the nerves.

Case Report: Variant B

This 50-year-old woman (Figs. 9.14–9.19) suffered from increasingly insecure gait, accompanied by cerebellar symptoms predominantly, on the right, discreet hypesthesia of the second and third trigeminal branches on the right, and pronounced hypacusia and taste impairment. Radiology revealed a giant petroclival meningioma of supra- as well as infratentorial extension that displaced and contorted the brain stem.

Surgery was carried out according to the modification of the supra-infratentorial approach described above. The patient was placed in a semisitting position; posterior subtemporal craniotomy was followed by lateral suboccip-

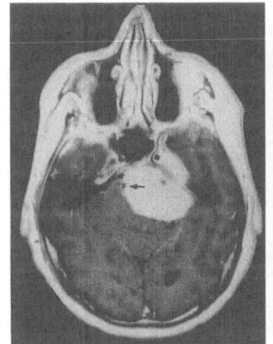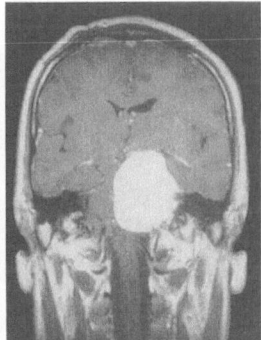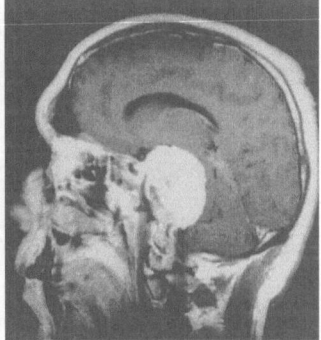

a   b, c

Fig. 9.14a–c. The preoperative MRI shows a meningioma of petroclivotentorial extension. The tumor has reached the lower cranial nerves, extends across the midline on the clivus and has shifted the basilar artery to the right (*arrow*). This tumor also involves the tentorium and Meckel's cavity

ital craniectomy. After mastoidectomy, labyrinthectomy was performed due to the advanced impairment of hearing. The facial nerve remained in its bony canal and formed the anterior limit of the partial petrosectomy. The dura was incised in the manner described above (Figs. 9.8 b, 9.10). The tentorium was split and then elevated with the temporal lobe. The tumor was removed in a piecemeal manner. In the course of dissection, the oculomotor and trochlear nerves were found to be cranially elevated. The trigeminal nerve, the facial nerve group, and the abducent nerve were encased in tumor. The caudal cranial nerves were displaced to the lower aspect of the tumor. For the sake of radicality, the abducent nerve on the right had to be resected.

Apart from the expected right abducent nerve palsy, mild left abducent paresis and right facial paresis occurred postoperatively.

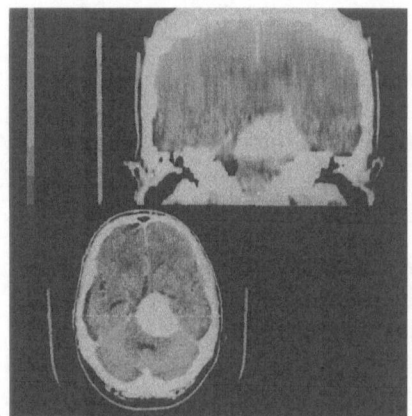
a

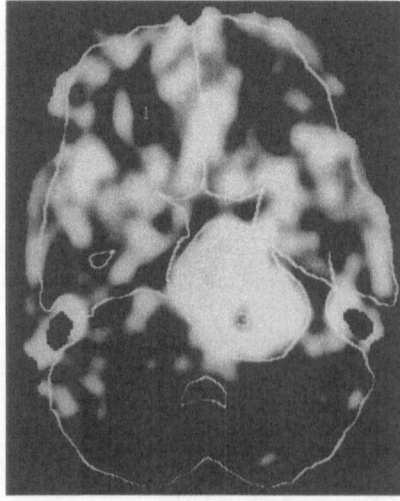
b

Fig. 9.15 a, b. CT: **a** coronal reconstruction and **b** typical high xenon uptake

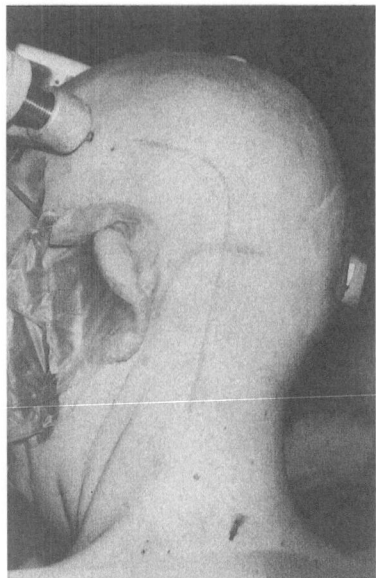

Fig. 9.16. Positioning the patient in a semisitting position with skin and localization of the transverse sinus

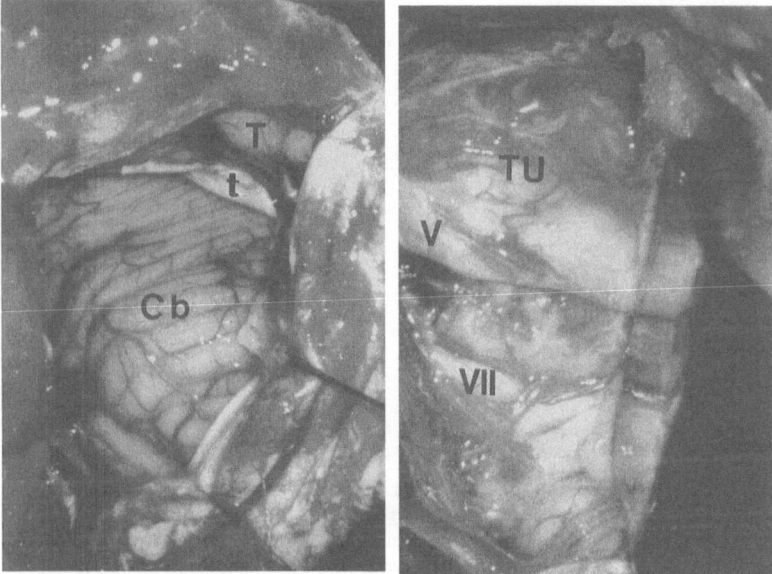

a   b

Fig. 9.17. a Combined presigmoidal petrous ridge approach from the left side with the dura opened anterior to the sigmoid sinus, which is fixed dorsally by sutures. The temporal dura is opened and the tentorium (*t*) has been split. *T,* Temporal lobe; *Cb,* cerebellum. **b** The approach into the posterior fossa with the tumor (*TU*) medial to the fifth (*V*) and seventh and eighth (*VII*) cranial nerves

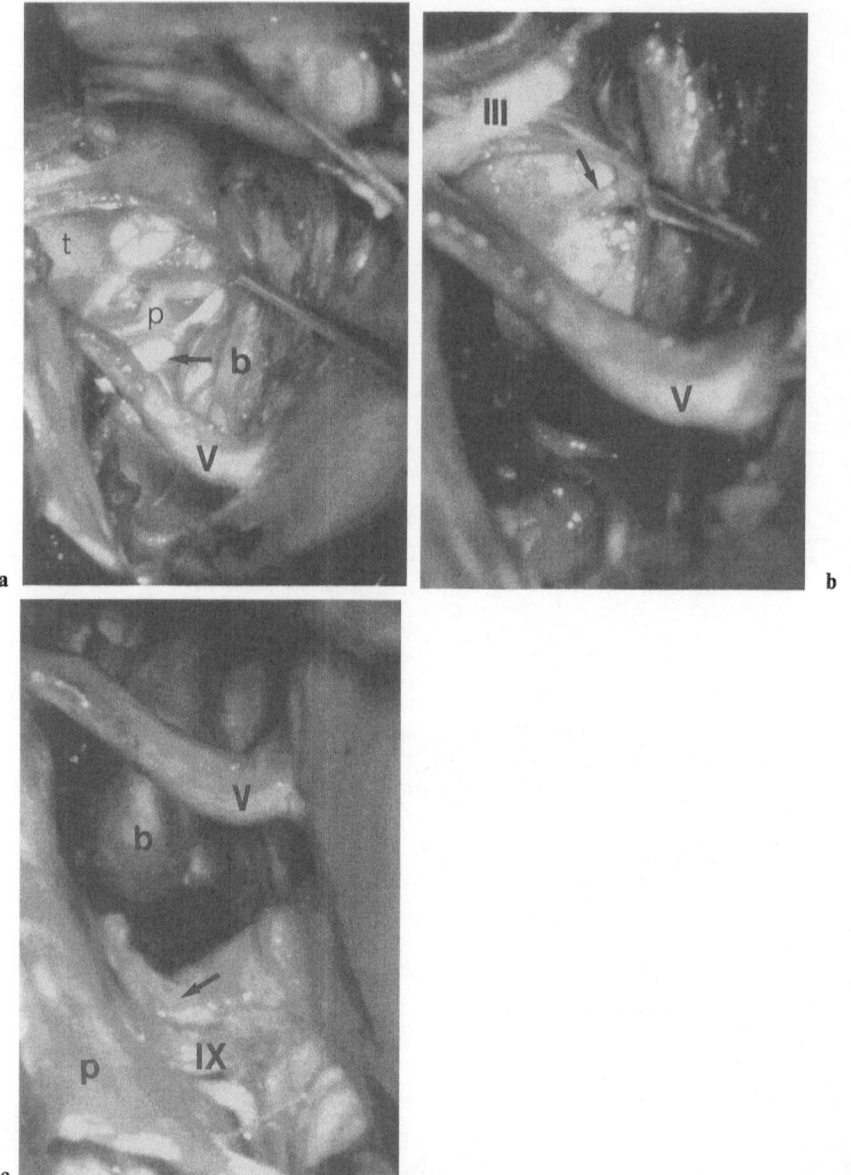

Fig. 9.18a–c. View into the prepontine space after tumor removal. **a** Oblique view into the interpeduncular fossa with the tuber cinereum (*t*). *p,* contralateral posterior communicating artery; *arrow* contralateral oculomotor nerve; *b,* basilar artery; *V,* ipsilateral trigeminal nerve. **b** Closer view of the contralateral abducent nerve (*arrow*) and oculomotor nerve (*III*). **c** After total tumor removal: the trigeminal nerve (*V*), the basilar artery (*b*), and the pons (*p*) are visible; the facial nerve group can be seen (*arrow*) at the level of the caudal cranial nerves (*IX*)

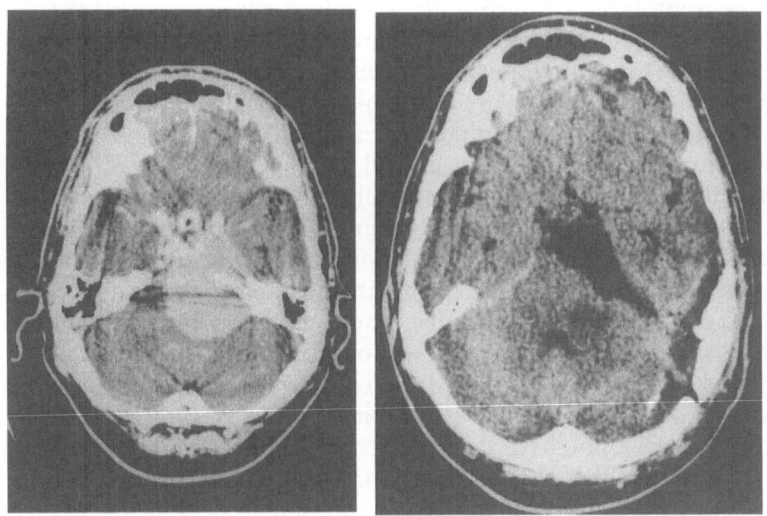

a   b

Fig. 9.19. a Pre- and **b** postoperative CT scans

Frequently Encountered Lesions

Meningiomas growing in the middle and posterior cranial fossae and addition-ally involving the tentorium constitute the majority of lesions operated on by this approach. The excision of the tentorium can be controlled from both sides and overview for the identification of brain stem structures is optimal. A further indication for the approach are meningiomas of the midclivus which do not particularly extend paramedian and do not contort the brain stem (Hakuba and Nishimura 1981; Symon 1982).

The approach also offers a better overview in cases of *complex giant aneurysms* of the basilar artery, for which the view afforded by the subtemporal approach alone would not suffice. It gives more space for the difficult operative maneu-vers that may be necessitated by the complexity of these aneurysms.

Cholesteatomas may extend across the tentorial notch to the middle cranial fossa. The approach is also suited for such lesions, transection of the transverse sinus may even be foregone in such cases, as the lesions are easy to remove.

Advantages and Disadvantages of the Approach

The approach lends itself well to cases in which large portions of tumor are situated in the middle and posterior cranial fossae and extirpation via a supra- or infratentorial approach alone is impossible. Whenever the tentorium is also involved, the approach of choice is the combined supra-infratentorial approach. It goes without saying that every neurosurgeon has special preferences for certain approaches, but at any rate one may keep one's options open for combined approaches for the sake of intraoperative flexibility, when, for example, performing lateral suboccipital craniectomy. After all, some problems cannot be predicted in even the most careful planning (Fig. 7.7).

In trepanations continued frontotemporally, structures from the optic chiasm to the dorsum sellae and from the dorsum sellae to the foramen magnum can be reached. Meningiomas of the posterior cranial fossa which invade the posterior part of the cavernous sinus along the dural entry areas of the third through sixth cranial nerves or along the venous pathways can be traced and the cavernous sinus can be opened in the way outlined above (see also Fig. 6.7).

Some authors, such as Symon or Malis, consider the combined supra-infratentorial approach best suited for prepontine space-occupying lesions. Dissection is toward the lesion from laterally, so that neither the cerebellum nor the temporal lobe is subjected to excessive retraction. When the most exterior part of the petrous bone has been drilled off, exposure in a cranial direction is further improved. Labyrinthectomy or petrosectomy for the sake of slightly better access is not justifiable if hearing is intact.

The major drawback of the combined supra-infratentorial approach is definitely the problem of venous drainage in transections of the transverse sinus. Although Symon (1982) mentions only one case in which fatal complications were caused by venous congestion, this is a risk that needs to be weighed carefully. Luyendijk (1976) therefore recommends protecting, the sigmoid sinus by a thin osseous lamella and not transecting it. This, however, causes the great advantage of a large-scale approach giving a good view to be lost.

The prerequisite for transection of the sigmoid sinus is angiographic evidence of the patency of the contralateral transverse and sigmoid sinuses and jugular vein (Sekhar et al. 1984). As in all subtemporal approaches, brain retraction basically entails a risk of temporal lobe epilepsy or hemorrhagic infarction of the temporal lobe; the risk is particularly high if drainage of the temporobasal veins is not ensured. We think that the technique described above, involving transection of the transverse sinus and retraction of the temporal lobe together with the transverse sinus, the tentorium, and the vein of Labbé, solves this problem, as the tentorium provides additional protection to the temporal lobe and temporal lobe drainage is provided.

In order to bypass most of the above-mentioned problems we looked for an improvement of the combined approach, which we found in variant B. As the approach follows the petrous ridge, we have suggested the name "presigmoidal petrous ridge approach".

In contrast to a conventional subtemporal approach, in this approach the temporal craniotomy is limited to 1–1.5 cm, which also helps to reduce the retraction of the temporal lobe to a minimum. Apart from the fact that cerebellar retraction is also minimized, this is what we consider the greatest advantage of this approach. Since only the superior petrosal sinus is ligated after the presigmoidal dura has been opened, the "venous problem" is substantially lower.

The decisive issue of this variant of approach is audition. If hearing is intact or there is a reasonable chance of later improvement the labyrinth must be preserved. This means that the approach may be confined to a rather small space and that extradural portions of tumor at the apex of the petrous bone cannot be reached.

In cases in which hearing has already been lost, much space can be gained by labyrinthectomy and possibly also by opening of the internal auditory canal. In such cases, the facial nerve, which is left in its canal, forms the limit of the petrous bone resection. Out of a total number of 24 petroclival meningiomas, 64% were extirpated with no operative fatalities and a low morbidity rate (Samii et al. 1989). Various approaches were used, but given the good results obtained using the presigmoidal petrous ridge approach, this modification will be our method of choice in the future.

Summary: Combined Subtemporal-Suboccipital Approach with Transection of the Sigmoid Sinus

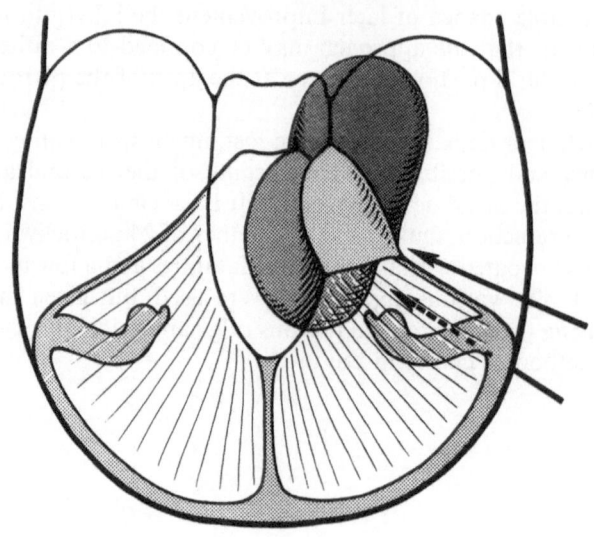

Indications: Huge petroclivotentorial lesions

Meningiomas – Epidermoids – Chordomas – Complex Basilar Aneurysms

Advantages: View from the optic nerve to foramen magnum
Extension possible in any direction
Wide exposure

Disadvantages: Problems of venous drainage
Temporal lobe risk

Limits: Intrapetrous internal carotid artery (?)
Cavernous sinus (?)
Seventh cranial nerve (?)

Summary: Presigmoidal Petrous Ridge Approach

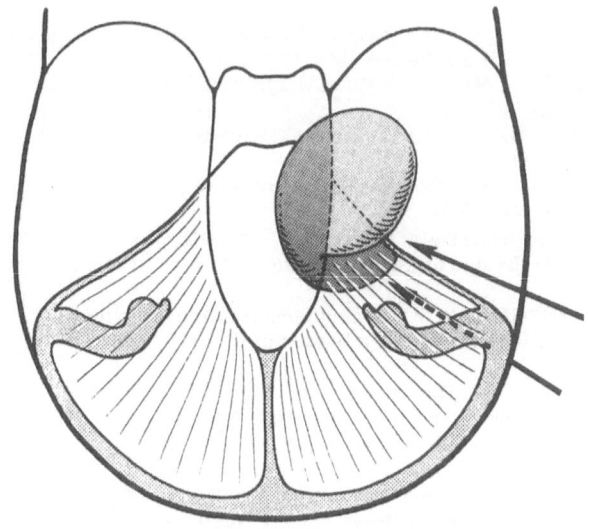

Indications: Petroclivotentorial lesions

Meningiomas – Trigeminal neurinomas – Epidermoids

Advantages: No venous problems
No temporal lobe problem
View across the midline

Disadvantages: Risk of hearing loss
CSF leak

Limits: Jugular bulb
Labyrinth (?)
Seventh cranial nerve

References

Al Mefty O (1987) Supraorbital-pterional approach to skull base lesions. Neurosurgery 21:474–477

Al Mefty O (1989) Surgery of the cranial base. In: Saleman M (ed) Foundations of neurolog-. ical surgery. Kluwer Academic, Boston

Al Mefty O, Fox JF, Smith RR (1988) Petrosal approach for petroclival meningiomas. Neurosurgery 22:510–517

Apuzzo ML, Weiss MH, Heiden J (1978) Transoral exposure of the atlanto axial region. Neurosurgery 3:201–207

Archer DJ, Young St, Uttley D (1987) Basilar aneurysms: a new transclival approach via maxillotomy. J Neurosurg 67:54–58

Arita N, Mori S, Sano M, Hayakawa T, Nakao K, Kanai N, Mogami H (1989) Surgical treatment of tumors in the anterior skull base using the transbasal approach. Neursurgery 24:379–384

Arnold H, Herrmann HD (1986) Skull base chordoma neurosurgical problems. In: Scheunemann H, Schürmann K, Helms J (eds) Tumors of the skull base. De Gruyter, Berlin, pp 209–216

Bartal AD, Heilbronn YD (1970) Transcervical removal of a clivus chordoma in a 2-year-old child. Reversal of quadriplegia and lumbar paralysis. Acta Neurochir (Wien) 23:127–133

Bailey P (1939) Concerning the technique of operations for acoustic neurinoma. Zentralbl Neurochir 4:1–5

Bergland R, Ray BS, Torack R (1968) Anatomical variations in the pituitary gland and adjacent structures in 225 human autopsy cases. J Neurosurg 28:93–99

Bisaria KK (1985) Anatomical variation of venous sinuses in the region of torcular Herophili. J Neurosurg 62:90–95

Black S, Ansbacher L (1984) Saccular aneurysm associated with segmental duplication of the basilar artery. J Neurosurg 61:1005–1008

Bogorodinsky DK (1936) Syndrome of craniospinal tumor. Tashkent Govt. Publ. (USSR):104

Böhler J (1982) Anterior stabilization for acute fracture and non union of the dens. J Bone Joint Surg (Am) 64:18–28

Bonnal J, Sedan R, Paillas JE (1961) Problèmes cliniques evolutifs et therapeutiques soulevés par les méningéomes envahissants de la base du crâne. Neurochirurgie 7:108–117

Bonnal J, Louis R, Combalbert A (1964) L'abord temporal transtentorial de l'angle ponto-cérébelleux et du clivus. Neurochirurgie 10:3–12

Bonnal J, Thibaut A, Brotchi A, Born J (1980) Invading meningeomas of the sphenoid ridge. J Neurosurg 53:587–599

Borchardt M (1905) Zur Operation der Tumore des Kleinhirnbrückenwinkels. Klin Wochenschr 42:1003–1035

Boskovic M, Savic V, Josifov J (1963) Über die Sinus petrosi und ihre Zuflüsse. Gegenbaurs Morphol Jahrb 104:420–429

Brackmann DE (1986) Otoneurosurgical procedures. In: May M (ed) The facial nerve. Thieme, New York, pp 589–617

Brihaye J, Hoffmann G, François J (1968) Les exophthalmies neurochirurgicales. Neurochirurgie 14:188–487

Busch W (1966) Beitrag zur Morphologie und Pathologie der Arteria basilaris. Arch Psychiatr Nervenkr 208:326–344

Castellano F, Ruggiero G (1953) Meningiomas of the posterior fossa. Acta Radiol [Suppl] (Stockh) 104:1–157

Chiari O (1912) Über eine Modifikation der Schloffer'schen Operation von Tumoren der Hypophyse. Wien Klin Wochenschr 25:5–6

Cloward RB (1958) The anterior approach for removal of ruptured cervical discs. J Neurosurg 15:602–627

Cloward RB (1972) Treatment of lesions of the cervical spine by the anterior cervical approach. In: Austin G (ed) The spinal cord, 2nd edn. Thomas, Springfield

Cloward RB, Passarelli P (1979) Removal of giant clival chordoma by anterior cervical approach. Surg Neurol 11:129–134

Cohen L, Macrae D (1962) Tumors in the region of the foramen magnum. J Neurosurg 19:462–469

Crabtree JA, Britton BH, Pierce MK (1976) Carcinoma of the external auditory canal. Laryngoscope 86:405–412

Crockard HA, Bradford R (1985) Transoral, transclival removal of a schwannoma anterior to the craniocervical junction. J Neurosurg 62:293–295

Crockard HA, Chandra N (1991) The transoral approach for the management of intradural lesions at the craniovertebral junction: Review of 7 cases. Neurosurgery 28:88–98

Cushing H (1912) The pituitary body and its disorders. Lippincott, Philadelphia

Cushing H (1914) Surgical experience with pituitary disorders. JAMA 63:1515–1525

Cushing H (1917) Tumors of the nervus acusticus and the syndrome of the cerebellopontine angle. Saunders, Philadelphia

Cushing H, Eisenhardt L (1938) Meningiomas. Thomas, Springfield

Dahlin DC, MacCarty CS (1952) Chordoma: a study of fifty-nine cases. Cancer 5:1170–1178

Dandy WE (1917) Exhibition of cases. Bull Johns Hopkins Hosp 28:96–100

Dandy WE (1925) An operation for the total removal of cerebellopontine (acoustic) tumors. Surg Gynecol Obstet 41:129–148

Dandy WE (1941) Results of removal of acoustic tumors by the unilateral approach. Arch Surg 42:1026–1033

Dandy A, Delcour J, Laine E (1963) Les méningiomas du clivus. Neurochirurgie 9:249–277

Dastur DK, Wadia NH, Desai AD (1965) Medullospinal compression due to atlanto-axial dislocation and sudden hematomyelia during decompression. Pathology, pathogenesis and clinical correlation. Brain 88:897–924

De Andrade JR, MacNab J (1969) Anterior occipito-cervical fusion using an extrapharyngeal exposure. J Bone Joint Surg (Am) 51:162–226

Decker RE, Malis LI (1970) Surgical approaches to midline lesions at the base of the skull: a review. Mt Sinai J Med (NY) 37:84–102

De Fries HO, Deeb ZE, Hudkins CP (1988) Arch Otolaryngol Head Neck Surg 144:766–769

Delandsheer JM, Caron JP, Jomin M (1977) Voie trans-buccopharyngée et malformations de la charnière cervico-occipitale. Neurochirurgie 23:276–281

Delgado TE, Garrido E, Harwick RD (1981) Labiomandibular, transoral approach to chordomas in the clivus and upper cervical spine. Neurosurgery 8:675–679

De Martel T, Guillaume J (1931) Surgical treatment of cerebral tumors. Technical considerations. Surg Gynecol Obstet 52:381–385

De Martel T, Thurel R (1936) Méningiome de la région de l'angle pontocérébelleux. Rev Oto-Neuro-Ophtalmol (Paris) 14:110–112

Dechaume J, Wertheimer P (1936) Les méningeomes rétro sellaires. Sud Med Chir 68:1072–1091

Denecke HJ (1966) Zur Chirurgie ausgedehnter Glomustumoren im Bereich des Foramen jugulare. Arch Otorhinolaryngol 187:656–662

Denecke HJ (1969) Surgery of extensive glomus jugulare tumors of the ear. Rev Laryngol 90:265–270

Denecke HJ (1978) Die Chirurgie ausgedehnter Tumoren des Felsenbeins und der Otobasis. Laryngol Rhinol 57:287–290

Denecke HJ (1980) Operative Korrektur des Schluckaktes und der Stimme bei einseitiger Vaguslähmung. In: Denecke HJ (ed) Die oto-rhino-laryngologischen Operationen im Mund- und Halsbereich. Springer, Berlin Heidelberg New York, p 678

Denker A (1921) Hypophysentumoren. Int Zentralbl Laryngold Rhinol 37:225

Derome P (1977a) La voie trans-bucco-pharyngée et la pathologie tumorale du clivus. Neurochirurgie 23:298–306

Derome P (1977b) The transbasal appraoch to tumors invading the base of the skull. In: Schmidek HH, Sweet WH (eds) Current techniques in operative neurosurgery. Grune and Stratton, New York, pp 223–245

Derome P, Caron JP, Hurth M (1977) Indications de la voie trans-bucco-pharyngée et malformations de la charnière cranio-vertébrale. Neurochirurgie 23:282–285

Derome PJ (1972) Les tumeurs sphéno-ethmoidales: possibilités d'exérèse et de réparation chirurgicales. Rapport de la Société de Neuro-Chirurgie de Langue Française. Neurochirurgie 15 [suppl 1]:1–164

Derome PJ, Guiot G (1979) Surgical approaches to sphenoidal and clival areas. In: Krayenbühl H (ed) Advances and technical standards in neurosurgery vol 6. Springer, Vienna New York, pp 101–136

Di Chiro G (1962) Angiographic patterns of cerebral convexity veins and superficial dural sinuses. Am J Roentgenol 87:308–321

Di Lorenzo N (1982) Craniovertebral junction malformations. Clinicoradiological findings, long term results and surgical indications in 63 cases. J Neurosurg 57:603–608

Dolenc VV (1983) Direct microsurgical repair of intracavernous vascular lesions. J Neurosurg 58:824–831

Dolenc VV (1985) A combined epi- and subdural direct approach to carotid ophthalmic artery aneurysms. J Neurosurg 62:667–672

Dolenc VV, Skrap M, Sustersic J, Surbec M, Morina A (1987a) A transcavernous-transsellar approach to basilar tip aneurysms. Br Neurosurg 1:251–259

Dolenc VV, Kregar T, Ferluga M, Fettich M, Morina A (1987b) Treatment of tumors invading the cavernous sinus. In: Dolenc VV (ed) The cavernous sinus. Springer, Vienna New York, pp 377–391

Dott NM (1963) Facial nerve reconstruction by graft bypassing the petrous bone. Arch Otolaryngol 78:426–428

Draf W, Samii M (1977) Otorhinolaryngo-neurochirurgische Probleme an der Schädelbasis. Laryngol Rhinol 56:1007

Draf W, Samii M (1982) Diagnostik und operative Strategie bei großen Glomustumoren der lateralen Schädelbasis. In: Majer H, Rieder CH (eds) Aktuelles in der Otolaryngologie. Thieme, Stuttgart, pp 61–70

Draf W, Samii M (1986) Malignant tumors of the paranasal sinuses. In. Scheunemann H, Schürmann K, Helms J (eds) Tumors of the skull base. De Gruyter, Berlin, pp 63–70

Drake CHG (1965) Surgical treatment of ruptured aneurysms of the basilar artery. Experience with 14 cases. J Neurosurg 23:457–473

Drake CHG (1969) The surgical treatment of vertebral basilar aneurysms. Clin Neurosurg 16:114–169

Drake CHG (1973) Management of aneurysms of posterior circulation. In: Youmans JR (ed) Neurological surgery, vol 2. Philadelphia, Saunders, pp 787–806

Eldridge MN, Smith RA (1967) Transoral fusion of odontoid fracture. Case report. J Neurosurg 27:462–465

Engel A (1975) Ursprungs- und Verlaufsvarietäten der ersten Ophthalmicastrecke. Medical dissertation, University of Würzburg

Falconer MA, Bailey IC, Duchen LN (1968) Surgical treatment of chordoma and chondroma of the skull base. J Neurosurg 29:261–275

Fang HSY, Ong GB (1962) Direct anterior approach to the upper cervical spine. J Bone Joint Surg [Am] 44A:1558–1604

Fang HSY, Ong GB, Hodgson AR (1964) Anterior spinal fusion. The operative approaches. Clin Orthop 35:16–33

Fay T (1930) The management of tumors of the posterior fossa by the transtentorial approach. Surg Clin North Am 10:1422–1499

Fein J (1910) Zur Operation der Hypophyse. Wien Klin Wochenschr 23:1035

Firbas W (1985) Makroskopische und mikroskopische Anatomie des statoakustischen Organs. In: Zenker W (ed) Makroskopische und mikroskopische Anatomie des Menschen, Benninghoff, vol 3. Urban und Schwarzenberg, Munich, pp 527–576

Fisch U (1978) Infratemporal fossa approach to tumors of the temporal bone and base of the skull. J Laryngol Otol 92:949–967

Fisch U (1977) Infratemporal fossa approach for extensive tumors of the temporal bone and base of skull. In: Silverstein H, Norell H (eds) Neurological surgery of the ear. Aesculapius, Birmingham, pp 34–53

Fisch U, Pillsbury HC (1979) Infratemporal fossa approach to lesions in the temporal bone and base of the skull. Arch Otolaryngol 105:99–107

Fox JL (1967) Obliteration of midline vertebral artery aneurysm via basilar craniotomy. J Neurosurg 26:406–412

Fox JL, Jerez A (1977) An unusual atlanto-axial dislocation. Case report. J Neurosurg 47:115–118

Fränkel J (1904) Contribution to the surgery of neurofibroma of the acoustic nerve. Ann Surg 40:293–319

Fujii K, Chambers SG, Rhoton AL (1979) Neurovascular relationships of the sphenoid sinus. J Neurosurg 50:31–39

Gacek RA, Goodman M (1977) Management of malignancy of the temporal bone. Laryngoscope 87:1622–1634

Gibo H, Lenkey C, Rhoton AL (1981) Microsurgical anatomy of the supraclinoid portion of the carotid artery. J Neurosurg 55:560–575

Gilsbach J, Seeger W (1988) Operative Zugänge nach anatomischen Gesichtspunkten. In: Hase H, Reulen H (eds) Die akute Raumforderung in der hinteren Schädelgrube. Überreuter, Vienna, pp 87–105

Glasscock ME (1983) Management of aneurysms of the petrous portion of the internal carotid artery by resection and primary anastomosis. Laryngoscope 93:1445–1453

Glasscock ME, Harris PF, Newsome G (1974) Glomus tumors: diagnosis and treatment. Laryngoscope 84:2006–2032

Glasscock ME, Miller GW, Drake FD, Kanok MM (1978) Surgery of the skull base. Laryngoscope 88:905–923

Glasscock ME, Pensak ML, Gulya AJ (1985) Surgery of the skull base. In: Rand RW (ed) Microneurosurgery. Mosby, St Louis, pp 421–452

Greenberg AD (1968) Atlanto axial dislocations. Brain 91:655–684

Greenberg AD, Scoville WB, Davey LM (1968) Transoral decompression of atlanto-axial dislocation due to odontoid hypoplasia. Report of two cases. J Neurosurg 28:266–269

Gudmundsson K, Rhoton AL, Rushton JG (1971) Detailed anatomy of the intracranial portion of the trigeminal nerve. J Neurosurg 35:592–600

Guidetti B, Spallone A (1980) Benign extramedullary tumors of the foramen magnum. Surg Neurol 13:9–17

Guidetti B, Spallone A (1988) Benign extramedullary tumors of the foramen magnum. In: Krayenbühl H (ed) Advances and technical standards in neurosurgery, vol 16. Springer, Vienna New York, pp 83–120

Guiot G (1958) Adenomas hypophysaires. Masson, Paris

Guiot G (1968) The rhinoseptal route for the removal of clivus chordomas. Johns Hopkins Med J 122:329–335

Guiot G, Derome P (1966) A propos les méningéomes en plaque du Ptérion le traitement chirurgical des méningéomes osseux hyperostosans. Ann Chir 20:19–20

Guiot G, Derome P (1976) Surgical problems of pituitary adenomas. In: Krayenbühl H (ed) Advances and technical standards in neurosurgery, vol 3. Springer, Vienna New York, pp 3–35

Guiot G, Thibaut B (1958) L'estirpation des adenomes hypophysaires par voie transsphenoidale. Neurochirurgia (Stuttg) 1:133–150

Guthkelch AN, Williams RG (1972) Anterior approach to recurrent chordomas of the clivus. J Neurosurg 36:670–672

Hakuba A (1986) The transpetrosal, transtentorial approach and its application in therapy of retrochiasmatic craniopharyngeomas. In: Samii M (ed) Surgery around the brainstem and third ventricle. Springer, Berlin Heidelberg New York, pp 396–404

Hakuba A, Nishimura S (1981) Total removal of clivus meningiomas and the operative results. Neurol Med Chir (Tokyo) 21(1):59–73

Hakuba A, Nishimura S, Tanaka K, Kishi H, Nakamura T (1977) Clivus meningiomas. Six cases of total removal. Neurol Med Chir (Tokyo) 17:63–77

Hardley MN, Martin NA, Spetzler RF, Sonntag VKH, Johnson PC (1988) Comparative transoral dural closure techniques: a canine model. Neurosurgery 22:392–397

Hardley MN, Spetzler RF (1986) The transoral surgical approach to the cranio-cervical junction. In: Samii M (ed) Surgery in and around the brainstem and third ventricle. Springer, Berlin Heidelberg New York, pp 467–475

Hardy J (1969a) Transsphenoidal microsurgery of normal and pathological pituitary. Clin Neurosurg 16:185–217

Hardy J (1969b) Microsurgery of the hypophysis: subnasal transsphenoidal approach with televised magnification and televised radiofluoroscopic control. In: Rand RW (ed) Microneurosurgery. Mosby, St Louis, pp 87–103

Hardy J (1971) Transsphenoidal hypophysectomy. Neurosurgical techniques. J Neurosurg 34:195–196

Hardy J (1977) L'abord transsphenoidal des tumeurs du clivus. Neurochirurgie 23:287–297

Hardy J, Wigser SM (1965) Transsphenoidal surgery of pituitary fossa tumors with televised radiofluoroscopic control. J Neurosurg 23:612–619

Harris FS, Rhoton AL (1976) Anatomy of the cavernous sinus, a microsurgical study. J Neurosurg 45:169–180

Hashi K, Hakuba A, Ikuno H (1976) A midline vertebral artery aneurysm operated via transclival approach. Neurol Surg (Jpn) 42:183–189

Hayakava T, Kamikawa K, Ohnishi T, Yoshimine T (1981) Transoral, transclival approach to aneurysms of the basilar artery. Neurol Med Chir (Tokyo) 21:477–484

Helms J (1981) Variations of the course of the facial nerve in the middle ear and mastoid. In: Samii M, Jannetta P (eds) The cranial nerves. Springer, Heidelberg Berlin New York, pp 391–393

Heros RC (1986) Lateral suboccipital approach for vertebral and vertebrobasilar lesions. J Neurosurg 64:559–562

Hieshima GB, Mehringer CM, Grinnell VS, Pribram HF, Tsai FY (1982) Embolization for tumors of the skull base. In: Brackmann DE (ed) Neurological surgery of the ear and skull base. Raven, New York, pp 391–394

Hirsch O (1909) Eine neue Methode der endonasalen Operation von Hypophysentumoren. Wien Med Wochenschr 59:636–637

Hirsch O (1910) Endonasal method of removal of hypophyseal tumors. JAMA 55:772–774

Hirsch O (1958) Hypophysentumore – ein Grenzgebiet. Acta Neurochir (Wien) 5:1–10

Hitselberger WE, House WF (1966) A combined approach to the cerebellopontine angle. A suboccipital-petrosal approach. Arch Otolaryngol 84:267–286

Hitselberger WE, House WF (1980) A warning regarding the sitting position for acoustic tumor surgery. Arch Otolaryngol 106:69

Hoffman WF, Wilson ChB (1979) Fenestrated basilar artery with an associated saccular aneurysm. J Neurosurg 50:262–264

House WF (1968) Surgical exposure of the internal auditory canal and its contents through the middle cranial fossa. Laryngoscope 71:1363–1385

House WF, Hitselberger WE (1969) The middle fossa approach for removal of small acoustic tumors. Acta Otolaryngol 67:413–427

House WF, Hitselberger WE (1976) The transcochlear approach to the skull base. Arch Otolaryngol 102:334–342

Jackson IT, Marsh WR, Bite U, Hide TA (1986) Craniofacial osteotomies to facilitate skull base tumor resection. Br J Plast Surg 39:153–160

Janecka IP, Shekar LN (1988) Anterior and anterolateral craniofacial resection. In: Janecka IP, Shekhar LN (eds) International symposium on cranial base surgery, Pittsburgh

Jenkins HA, Fisch U (1981) Glomus tumors of the temporal region: technique and surgical resection. Arch Otolaryngol 107:209–214

Jomin M, Bouasakao U (1977) Technique chirurgicale de la voie transbuccale. Neurochirurgie 23:259–264

Kawase T, Toya S, Shiobara R, Mine T (1985) Transpetrosal approach for aneurysms of the lower basilar artery. J Neurosurg 63:857–861

Kawase T, Toya S, Shiobara R, Kimura C, Nakajima H (1987) Skull base approaches for meningeomas invading the cavernous sinus. In: Dolenc VV (ed) The cavernous sinus. Springer, Vienna New York, pp 346–354

Kaye AH, Hahn JF, Kinney SE, Hardy J, Bay JW (1984) Jugular foramen schwannomas. J Neurosurg 60:1045

Kerber ChB, Manke W (1983) Trigeminal artery to cavernous sinus fistula treated by balloon occlusion. J Neurosurg 58:611–613

Key A, Retzius G (1875) Studien in der Anatomie des Nervensystems und des Bindegewebes, Bd 1. Norstad, Stockholm

Kline LB, Glaser JS (1981) Bilateral abducens nerve palsies from clivus chordoma. Ann Opthalmol 13:705–707

Knosp E, Müller G, Perneczky A (1987a) The blood supply of the cranial nerves in the lateral wall of the cavernous sinus. In: Dolenc VV (edn) The cavernous sinus. Springer, Vienna New York, pp 67–80

Knosp E, Müller G, Perneczky A (1987b) Anatomical remarks on the fetal cavernous sinus and on the veins of the middle cranial fossa. In: Dolenc VV (ed) Springer, Vienna New York, pp 104–116

Knosp E, Müller G, Perneczky A (1988a) The paraclinoid carotid artery: anatomical aspects of a microneurosurgical approach. Neurosurgery 22:896–901

Knosp E, Krisch K, Schidbauer M, Budka H (1988b) Immunologischer Hormonnachweis bei Hypophysenadenomen: Korrelation von Serumhormonbefunden mit immuncytochemischem Hormonbefund am Tumorschnitt. Wien Klin Wochenschr 100:322–325

Knosp E, Tschabitscher M, Matula CH, Koos WTH (1991) Modifications of temporal approaches: anatomical aspects of a microneurosurgical approach. Acta Neurochir (Wien) (in press)

Koos WTH, Perneczky A (1982) Neurochirurgie des Acusticusneurinoms. In: Majer E, Rieder CH (eds) Aktuelles in der Otorhinolaryngologie. Thieme, Stuttgart, pp 25–29

Koos WTH, Perneczky A (1985) The suboccipital approach to the acoustic neurinoma with emphasis on preservation of facial nerve and cochlear nerve function. In: Rand RW (ed) Microneurosurgery. Mosby, St Louis, pp 335–365

Koos WTH, Böck F, Spetzler R (1976) Clinical microneurosurgery. Thieme, Stuttgart

Koos WTH, Spetzler RF, Pendl G, Perneczky A, Lang J (1985) Color atlas of microneurosurgery. Thieme, Stuttgart

Krause F (1904) Zur Freilegung der hinteren Felsenbeinfläche und des Kleinhirns. Beitr Klin Chir 37:728–764

Krayenbühl H, Yaşargil MG (1975a) Chondromas. Prog Neurol Surg 6:435–463

Krayenbühl H, Yaşargil MG (1975b) Cranial chordomas. Prog Neurol Surg 6:380–434

Krayenbühl H, Yaşargil MG (1979) Zerebrale Angiographie für Klinik und Praxis. Thieme, Stuttgart

Kumar A, Fisch U (1983) The infratemporal fossa approach for lesions of the skull base. In: Krayenbühl H (ed) Advances and technical standards in neurosurgery, vol 10. Springer, Vienna New York, pp 187–220

Kurze T (1979) Microsurgery of the posterior fossa. Clin Neurosurg 26:463–478

Laine E, Jomin M (1977) Indications et possibilités de la voie trans-bucco-pharyngée en présence d'un aneurysme du confluent vertébro-basilaire. Neurochirurgie 23:307–314

Lang J (1975) Über Pori durales der Hirnnerven III, IV und VI. Verh Anat Ges 69:785–791

Lang J (1981) Klinische Anatomie des Kopfes. Springer, Berlin Heidelberg New York

Lang J (1985a) Über extradurale Ursprünge der A. cerebelli inferior posterior (PICA) und deren klinische Bedeutung. Neurochirurgie 28:183–187

Lang J (1985b) Anatomy of the midline. Koos WTH, Pendl G (eds) Lesions of the cerebral midline. Acta Neurochir [Suppl] (Wien) 35:6–22

Lang J (1986a) Craniocervical region, osteology and articulations. Neuro-Orthop 1:67–92

Lang J (1986b) Craniocervical region, central nervous system and envelopes. Neuroorthop 2:1-15

Lang J (1986c) Craniocervical region, blood vessels. Neuroorthop 2:55-69

Lang J (1987) Craniocervical region, surgical anatomy. Neuroorthop 3:1-26

Lang J, Schäfer K (1976) Über Ursprung und Versorgungsgebiete der intracavernösen Strecke der A. carotis interna. Gegenbaurs Morphol Jahrb 122:182-202

Lanz T v., Wachsmuth W (Hrsg) (1979) Praktische Anatomie, Bd 1, Teil 1. Springer, Berlin Heidelberg New York

Lasjaunias P (1981) Craniofacial and upper cervical arteries. Williams and Wilkins, Baltimore

Lasjaunias P, Moret J, Mink J (1977) The anatomy of the inferior lateral trunk of the internal carotid artery. Neuroradiology 13:215-220

Lautenschläger A (1929) Die prämaxilläre Hypophysenoperation. Chirurgie 1:30-33

Laws ER, Kern EB (1976) Complications of transsphenoidal surgery. Clin Neurosurg 23:401-416

Lesoin F, Jomin M, Pellerin P (1986) Transclival, transcervical approach to the upper cervical spine and clivus. Acta Neurochir (Wien) 80:100-104

Lewis JS (1983) Surgical management of tumors of the middle ear and mastoid. J Laryngol Otol 97:299-311

Luyendijk W (1976) Operative approach to the posterior fossa. In: Krayenbühl H (ed) Advances and technical standards in neurosurgery, vol 3. Springer, Vienna New York, pp 81-101

Malis LL (1975) Microsurgical treatment of acoustic neurinomas. In: Handa H (ed) Microneurosurgery. Igacu-Shoin, Tokyo

Martin RG, Grand JL, Peace D, Theiss C, Rhoton AL (1980) Microsurgical relationship of the anterior inferior cerebellar artery and the facial-vestibulocochlear nerve complex. Neurosurgery 6:483-507

Marx H (1913) Zur Chirurgie der Kleinhirnbrückenwinkeltumore. Mitt Grenzgeb Med Chir 26:117-134

Matricali B, Dulke M (1981) Aneurysm of fenestrated basilar artery. Surg Neurol 15:189-191

May M (1986) Anatomy of the facial nerve for the clinician. In: May M (ed) The facial nerve. Thieme, New York, pp 21-62

Mayberg M, Symon L (1986) Meningeomas of the clivus and apical petrous bone: report of 35 cases. J Neurosurg 65:160

Mazzoni A (1969) Internal auditory canal: arterial relations at the porus acusticus. Ann Otol Rhinol Laryngol 78:797-814

Menezes AH, VanGilder JC, Graf CJ, McDonnell DE (1980) Craniocervical abnormalities. A comprehensive surgical approach. J Neurosurg 53:444-455

Menezes AH, VanGilder JC, Clark CR, El Khoury G (1985) Odontoid migration in rheumatoid arthritis. J Neurosurg 63:500-509

Mitsushima T, Rhoton AL, De Oliveira E, Peace D (1983) Microsurgical anatomy of the veins of the posterior fossa. J Neurosurg 59:63-105

Mitterwallner FV (1955) Variationsstatistische Untersuchungen an den basalen Hirngefäßen. Acta Anat (Basel) 24:51-88

Morita A, Fukushima T, Miyazaki S, Shimitsu T, Atsuchi M (1989) Tic douloureux caused by primitive trigeminal artery or its variant. J Neurosurg 70:415-419

Morrison AW, King TT (1973) Experiences with a translabyrinthine-transtentorial approach to the cerebellopontine angle. J Neurosurg 38:382-390

Mullan S, Naunton R, Hekmat-Panach J, Vailati G (1966) The use of an anterior approach to ventrally placed tumors in the foramen magnum and vertebral column. J Neurosurg 24:536-543

Naffziger HC (1928) Brain surgery with special reference to exposure of the brainstem and posterior fossa. Surg Gynecol 46:240-248

Naffziger HC (1948) Exophthalmos. Some principles of surgical management from the neurosurgical aspect. Am J Surg 75:25-41

Nagashima CH, Iwasaki T, Okada K, Sakaguchi A (1979) Reconstruction of the atlas and axis with wire and acrylic after metastatic destruction. J Neurosurg 50:668–673

Nathan H, Quaknine G, Kosary IZ (1974) The abducens nerve. J Neurosurg 41:561–566

Norell H (1982) The sitting position for surgery: Techniques and pitfalls. In: Brackmann DE (ed) Neurological surgery of the ear and skull base. Raven, New York, pp 213–219

Oka K, Rhoton AL, Barry M, Rodriguez R (1985) Microsurgical anatomy of the superficial veins of the cerebrum. Neurosurgery 17:711–748

Olivecrona H (1967) Surgical treatment of intracranial tumors. Handbuch der Neurochirurgie vol IV. Springer, Berlin Heidelberg New York

Ono M, Rhoton AL, Barry M (1984) Microsurgical anatomy of the region of the tentorial incisure. J Neurosurg 60:365–400

Padget D (1948) The development of the cranial arteries in the human embryo. Contributions to embryology 212:257–261

Padget D (1956) The cranial verous system in man in reference to development, adult configuration and relation to the arteries. Am J Anat 98:307–353

Pariser SC (1977) The middle cranial fossa approach to the internal auditory canal. An anatomical study stressing critical distances between surgical landmarks. Laryngoscope 87 [Suppl 4]:1–20

Parkinson D (1964) Collateral circulation of the cavernous carotid artery: anatomy. Can J Surg 7:251–268

Parkinson D (1965) A surgical approach to the cavernous portion of the carotid artery: anatomical studies and case report. J Neurosurg 23:474–483

Parkinson D (1973) Carotid cavernous fistula: direct repair with preservation of the carotid artery. J Neurosurg 38:99

Parkinson D, Schields CB (1974) Persistant trigeminal artery: its relationship to the normal branches of the cavernous sinus. J Neurosurg 40:245–248

Parson H, Lewis JS (1954) Subtotal resection of the temporal bone for cancer of the ear. Cancer 7:995–1001

Pasztor E (1985) Transoral approach for epidural craniocervical pathological processes. In: Krayenbühl H (ed) Advances and technical standards in neurosurgery, vol 12. Springer, Vienna New York, pp 125–170

Pasztor E, Vajda J, Piffko P, Horvath M (1980) Transoral surgery in basilar impression. Surg Neurol 14:473–476

Pasztor E, Vajda J, Piffko P, Horvath M, Gador I (1984) Transoral surgery for craniocervical space-occupying processes. J Neurosurg 60:276–281

Paullus WP, Paid TG, Rhoton AL (1977) Microsurgical exposure of the petrous portion of the carotid artery. J Neurosurg 47:713–726

Pedroza A, Dujovny M, Ausmanj et al. (1986) Microvascular anatomy of the interpeduncular fossa. J Neurosurg 64:484–493

Peerless SJ, Drake CHC (1982) Aneurysms of the posterior circulation. In: Youmans JR (ed) Neurological surgery. Saunders, Philadelphia, pp 1715–1764

Perneczky A (1979) Die Arteria cerebelli inferior anterior. Anatomie, Klinik, Microneurochirurgie. Acta Chir Austriaca [Suppl] 29:1–42

Perneczky A (1980) The blood supply of acoustic neurinomas. Acta Neurochir (Wien) 52:209–218

Perneczky A (1986) The posterolateral approach to the foramen magnum. In: Samii M (ed) Surgery in and around the brain stem and the third ventricle. Springer, Berlin Heidelberg New York, pp 460–466

Perneczky A, Knosp E (1986) Intracavernous connective tissue cover of the internal carotid artery. Anatomy and surgery. In: Scheunemann H, Schürmann K, Helms J (eds) Tumors of the skull base. De Gruyter, Berlin, pp 171–175

Perneczky A, Knosp E, Vorkapic P, Czech TH (1985) Direct surgical approach to infraclinoidal aneurysms. Acta Neurochir (Wien) 76:36–44

Perneczky A, Knosp E, Czech TH (1987) Para- and infraclinoid aneurysms. Anatomy, surgical technique and report on 22 cases. In: Dolenc VV (ed) The cavernous sinus. Springer, Vienna New York, pp 252–271

Perneczky A, Knosp E, Matula CH (1988) Cavernous sinus surgery. Approach through the lateral wall. Acta Neurochir (Wien) 92:76–82

Pertuiset H (1974) Supratentorial craniotomy. In: Krayenbühl H (ed) Advances and technical standards in neurosurgery, vol 1. Springer, Vienna New York, pp 144–171

Pertuiset H, Beciric T (1955) La voie intracranienne dans l'exérèse du prolongement nasal des meningiomes olfactifs. Press Méd 63:615–618

Pertuiset B, Guillaumat L, Pialoux P, Hirsch JF (1958) Méningeome cranio-facial (temporoorbito-jugual) opéré en trois temps. Presse Med 66:1863–1865

Pitelli SD, Almeida GG, Nakagawa EJ, Marchese AJ, Cabral ND (1986) Basilar aneurysm surgery: the subtemporal approach with section of the zygomatic arch. Neurosurgery 18:125–128

Pool JL (1966) Suboccipital surgery for acoustic neurinomas: advantages and disadvantages. J Neurosurg 24:483

Poppen AL, King AB (1952) Chordoma: experience with 13 cases. J Neurosurg 9:139–163

Putz R (1975) Zur Manifestation der hypochordalen Spangen im cranio-cervicalen Grenzgebiet beim Menschen. Anat Anz 137:65

Raffel C, Wright DC, Gutin PH, Wilson CB (1985) Cranial chordomas: clinical presentation and results of operative and radiation therapy in twenty-six patients. Neurosurgery 17:703–710

Rand RW, Di Tullio M (1985) The Rand-Kurze suboccipital transmeatal operation for acoustic neuromas. In: Rand RW (ed) Microneurosurgery. Mosby, St Louis, pp 265–301

Rand RW, Kurze T (1965) Microneurosurgical resection of acoustic tumors by transmeatal posterior fossa approach. Bull Los Ang Neurol Soc 30:17–20

Renn WH, Rhoton AL (1975) Microsurgical anatomy of the sellar region. J Neurosurg 43:288

Resch KDM (1990) Zugangsanalyse und Zugangsdesign des transoral-pharyngealen Weges zum Hirnstamm. Thesis Med. fakult. Heidelberg

Rhoton AL (1968) N. intermedius. J Neurosurg 29:609–718

Rhoton AL, Pulec JL, Hall MG, Boyd AS (1967) Absence of bone over the geniculate ganglion. J Neurosurg 28:48–53

Rhoton AL, Hardy DG, Chambers SM (1979) Microsurgical anatomy and dissection of the sphenoid bone, cavernous sinus and sella area. Surg Neurol 12:63–104

Richling B (1982) Homologous controlled viscosity fibrin for endovascular embolization, part I. Experimental development of the medius. Acta Neurochir (Wien) 62:159–170

Rougerie J, Guiot G, Bouche J, Trigo JC (1967) Les voies d'abord du chordome du clivus. Neurochirurgie 13:559–570

Saeki N, Rhoton AL (1977) Microsurgical anatomy of the upper basilar artery and the posterior circle of Willis. J Neurosurg 46:563–578

Saito I (1978) Clipping of vertebrobasilar aneurysm by the transclival approach. Neurol Med Chir (Tokyo) 18:175

Samii M (1981) Preservation and reconstruction of the facial nerve in the cerebellopontine angle. In: Samii M, Jannetta P (eds) The cranial nerves. Springer, Berlin Heidelberg New York, pp 438–450

Samii M (1985) Microneurosurgery of acoustic neurinomas with special emphasis on preservation of seventh and eighth cranial nerve and the scope of facial nerve grafting. In: Rand RW (ed) Microneurosurgery. Mosby, St Louis, pp 366–388

Samii M (1986) Neurosurgical aspects of processes at the tentorial margin. In: Samii M (ed) Surgery in and around the brain stem and the third ventricle. Springer, Berlin Heidelberg New York, pp 416–443

Samii M (1991) Reconstruction of V (in press)

Samii M, Draf W (1986) The diagnosis and operative strategy of large glomus tumors. In: Scheunemann H, Schürmann K, Helms J (eds) Tumors of the skull base. De Gruyter, Berlin, pp 237–244

Samii M, Draf W (1989) Surgery of the skull base. Springer, Berlin Heidelberg New York

Samii M, von Wild K (1981) Operative treatment of lesions in the region of the tentorial notch. Neurosurg Rev 4:3–10

Samii M, Ammirati M, Mahram A, Bini W, Sephernia A (1989) Surgery of petroclival meningeomas: report of 24 cases. Neurosurgery 24:12–17

Samson DS, Hodosh RM, Clark WR (1978) Microsurgical evaluation of the pterional approach to aneurysms of the distal basilar circulation. Neurosurgery 3:135–141

Sano K, Jimbo M, Saito I (1966) Vertebrobasilar aneurysm with special reference to the transpharyngeal transclival approach to the basilar artery aneurysm. Brain Nerve 18:1197–1203

Schisano G, Tovi D (1962) Clivus chordomas. Neurochirurgia (Stuttg) 5:99–120

Schloffer H (1906) Zur Frage der Operationen an der Hypophyse. Beitr Klin Chir 50:767–817

Schloffer H (1907) Erfolgreiche Operationen eines Hypophysentumors auf nasalem Wege. Wien Klin Wochenschr 20:621–624

Schmelzle R, Harms JA (1986) A new transoral approach for surgery of malformations of the craniocervical junction. Report of 42 cases (abstract). Skull base study group, 3rd international meeting, Monte Carlo

Schürmann K, Voth D (1972) Die Bedeutung der transfrontalen Orbitotomie für die operative Behandlung der intraorbitalen raumfordernden Prozesse. Adv Ophthalmol 25:188–239

Seeger W (1980) Microsurgery of the brain, part 2. Springer, Vienna New York

Sekhar LN, Moller AM (1986) Operative excision of tumors involving the cavernous sinus. J Neurosurg 64:879–890

Sekhar LN, Samii M (1986) Petroclival and medial tentorial meningeomas. In: Scheunemann H, Schürmann K, Helms J (eds) Tumors of the skull base. De Gruyter, Berlin, pp 141–158

Sekhar LN, Jannetta PJ, Maroon IC (1984) Tentorial meningeomas. Surgical management and results. Neurosurgery 14:268–275

Shiu PHC, Hanafee WN, Wilson GH, Rand RW (1968) Cavernous sinus venography. Am J Roentgenol 104:57–62

Solomon RA, Stein BM (1988) Surgical approaches to aneurysms of the vertebral and basilar arteries. Neurosurgery 23:203–208

Southwick WO, Robinson RA (1957) Surgical approaches to the vertebral bodies in the cervical and lumbar region. J Bone Joint Surg 39:631–644

Spetzler RF, Selman WR, Nasa CL (1979) Transoral microsurgical odontoid resection and spinal cord monitoring. Spine 4:506–510

Spiess G (1911) Tumor der Hypophysengegend auf transnasalem Weg erfolgreich operiert. Muench Med Wochenschr 58:2503

Stein BM, Leeds NE, Taveras IM, Pool JL (1963) Meningioma of the foramen magnum. J Neurosurg 20:740–751

Stevenson GC, Stoney RJ, Perkins RK, Adams JE (1966) A transcervical transclival approach to the ventral surface of the brain stem for removal of a clivus chordoma. J Neurosurg 64:544–551

Sugita K (1985) Microneurosurgical atlas. Springer, Berlin Heidelberg New York

Sugita K, Kobayashi S, Shintani A, Mutsuga N (1979) Microneurosurgery for aneurysms of the basilar artery. J Neurosurg 51:615–620

Sugita K, Kobayashi S, Takeae T, Tada T, Tanaka Y (1987) Aneurysms of the basilar trunk. J Neurosurg 66:500–506

Sukoff MH, Kadin MM, Moran T (1972) Transoral decompression for myelopathy caused by rheumatoid arthritis of the cervical spine: case report. J Neurosurg 37:493–497

Sunderland S (1945) The arterial relations of the internal auditory meatus. Brain 68:23–27

Symon L (1982) Surgical approaches to the tentorial hiatus. In: Krayenbühl H (ed) Advances and technical standards in neurosurgery, vol 9. Springer, Vienna New York, pp 70–112

Taptas NJ (1982) The so-called cavernous sinus: a review of the controversy and its implications for neurosurgeons. Neurosurgery 11:712–717

Tessier P (1973) The conjunctival approach to the orbital floor and maxilla in congenital malformation and trauma. J Maxillofac Surg 1:3–8

Tessier P, Guiot G, Rougerie J (1967) Osteotomies cranio-naso-orbito-faciales hyperteloris-me. Ann Chir Plast 12:103–118

Tessier P, Guiot G, Derome PJ (1973) Orbital hypertelorism. II: Definite treatment of orbital hypertelorism by craniofacial or by extracranial osteotomies. Scand J Plast Reconstr Surg 7:39–58

Tiefental H (1920) Technik der Hypophysenoperation. Muench Med Wochenschr 67:794

Torklus D, Gehle W (1987) Die obere Halswirbelsäule. Thieme, Stuttgart

Tschabitscher M, Höcker K (1991) Der Nervus intermedius an seiner Ein- bzw. Aus-trittsstelle am Hirnstamm. Neurochirurgie (im Druck)

Tulleken CA, Luiten ML (1986) The basilar artery bifurcation in situ, approached via the sylvian route. An anatomical study in human cadavers. Acta Neurochir 80:109–115

Tschabischer M, Weber MW, Georgopoulos M (1990) The persistant trigeminal artery and it's topographical relations. Acta Anat 138:84–88

Tschabitscher M, Perneczky A (1974) Über Beziehungen von Kleinhirnarterien zum Meatus acusticus internus. Acta Anat 88:231–244

Tschabischer M, Weber MW, Georgopoulos M (1990) The persistant trigeminal artery and it's topographical relations. Acta Anat 138:84–88

Umansky F, Nathan H (1982) The lateral wall of the cavernous sinus with special reference to the nerves related to it. J Neurosurg 56:228–235

Umansky F, Elidan J, Valarezo A (1991) Dorello's canal: a microanatomical study. J Neuro-surg 75:294–298

Unterberger S (1958) Zur Versorgung frontobasaler Verletzungen. Arch Otolaryngol 172:463

Van Alyea OE (1941) Sphenoid sinus: anatomical study with consideration of the clinical significance of structural characteristics of the sphenoid sinus. Arch Otolaryngol 34:225–253

Van Buren JM, Omnaya AK, Ketchan MD (1968) Ten years experience with radical com-bined craniofacial resection of malignant tumors of the paranasal sinuses. J Neurosurg 28:341–350

Van Gilder JC, Menezes AHJ (1977) Cranio-vertebral abnormalities and their treatment. In: Schmidek HH, Sweet WH (eds) Current techniques in operative neurosurgery. Grune and Stratton, New York, pp 1221–1235

Verbiest H (1968) A lateral approach to the cervical spine, technique and indications. J Neurosurg 28:191–203

Verbiest H (1970) La chirurgie antérieure et latérale du rachis cervical. Neurochirurgie 16:1–121

Verbiest H (1973) The lateral approach to the cervical spine. Clin Neurosurg 20:295–305

Verbiest H (1977) Indications et possibilités de la voie transbucco-pharyngée. Neurochirurgie 23:513–516

Verbrugghen A (1952) Paragasserian tumors. J Neurosurg 9:451–460

Van Hayek H (1927) Untersuchungen über Epistropheus, Atlas und Hinterhauptsbein. Morphol Jahrb 58:269–347

Vorkapic P, Perneczky A, Tschabitscher M, Knosp E, Flohr A (1985) Transylvian approach to the tentorial hiatus – anatomical remarks on the microsurgical exposure. Zentralbl Neurochir 46:2–10

Wackenheim A (1985) Hypoplasia of the basi-occipital bone and persistance of the spheno-occipital synchondrosis in a patient with transitory supplementary fissure of the basi occipital. Neuroradiology 27:226–231

Weiss MH (1987) Transnasal transsphenoidal approach. In: Apuzzo ML (ed) Surgery of the third ventricle. William and Wilkins, Baltimore, pp 476–494

White RJ, Albin MS (1962) The technique and results of ligation of the basilar artery in monkeys. J Surg Res 2:15–18

Wissinger JP, Danoff D, Wisiol ES, French LA (1967) Repair of an aneurysm of the basilar artery by a transclival approach: case report. J Neurosurg 26:417–419

Wood BG, Sadar ES, Levine HL, Dohn DF, Tucker HM (1980) Surgical problems of the base of the skull. An interdisciplinary approach. Arch Otolaryngol 106:1–5

Yaşargil MG (1969) Microsurgery. Applied to neurosurgery. Academic/Thieme, Stuttgart

Yaşargil MG (1970) Intracranial microsurgery. Clin Neurosurg 17:250–255

Yaşargil MG (1976) Suboccipitale-transmeatale mikrochirurgische Exstirpation des Acusticusneurinoms. In: Naumann H (ed) Kopf- und Halschirurgie, Vol. 3. Thieme, Stuttgart, pp 545–587

Yaşargil MG (1984a) Microneurosurgery, vol 1. Thieme, Stuttgart

Yaşargil MG (1984b) Microneurosurgery, vol 2. Thieme, Stuttgart

Yaşargil MG, Fox JL, Ray MW (1975) The operative approach to aneurysms of the anterior communicating artery. In: Krayenbühl H (ed) Advances and technical standards in neurosurgery, vol 2. Springer, Vienna New York, pp 113–170

Yaşargil MG, Antic J, Laciga R, Jain KK, Hodosch RM, Smith RD (1976) Microsurgical pterional approach to aneurysms of the basilar bifurcation. Surg Neurol 6:83–91

Yaşargil MG, Kasdaglin K, Jain KK, Weber HP (1976) Anatomical observations of the subarachnoid cisterns of the brain during surgery. J Neurosurg 44:298–302

Yaşargil MG, Smith RD, Gasser JC (1977) Microsurgical approach to acoustic neurinomas. In: Krayenbühl, H (ed) Advances and technical standards in neurosurgery, vol 4. Springer, Vienna New York, pp 93–122

Yaşargil MG, Mortara RW, Curcic M (1980) Meningiomas of basal posterior cranial fossa. In: Krayenbühl H (ed) Advances and technical standards in neurosurgery, vol 7. Springer, Vienna New York, pp 4–115

Yaşargil MG, Reichman MW, Kubik ST (1987) Preservation of the fronto-temporal branch of the facial nerve using the interfascial temporalis flap for pterional craniotomy (technical article). J Neurosurg 67:463–467

Zeal AA, Rhoton AL (1978) Microsurgical anatomy of the posterior cerebral artery. J Neurosurg 48:534–559

Zoltan L (1974) Die Tumoren im Foramen occipitale magnum. Acta Neurochir (Wien) 30:217–225

Zoltan L, Fenyes L (1960) Stereotactic diagnosis and radioactive treatment in a case of spheno-occipital chordoma. J Neurosurg 17:888–900

Subject Index

M. Samii, Hannover Medical School; **M. Ammirati,** University of California, Los Angeles, CA

Surgery of Skull Base Meningiomas

**With a Chapter on Pathology
by G. F. Walter**

1992. Approx. 150 pp. 120 figs. in 245 sep. illus.
Hardcover ISBN 3-540-54016-4

This book sets the new standard on the surgical treatment of skull base meningiomas. What can be expected from surgery, when and where to be extremely aggressive and equally, or more important, when to stop. This book distills ten years' experience accrued treating more than 200 patients with skull base meningiomas at a single institution, thus conveying the authors' philosophy on the treatment of these lesions. Techniques and indications continuously improved over a decade are clearly presented and discussed and made available to the reader.

Springer-Verlag
Berlin
Heidelberg
New York
London
Paris
Tokyo
Hong Kong
Barcelona
Budapest

M. Samii, Hannover Medical School,
FRG (Ed.)

Surgery of the Sellar Region and Paranasal Sinuses

1991. XXII, 583 pp. 350 figs. 77 tabs.
Hardcover ISBN 3-540-53697-3

The challenging and interdisciplinary subject
of this book, the surgery of the sellar region and
paranasal sinuses, is considered and analysed from
different viewpoints, bringing together the concept
of the pathology, diagnostic procedures and thera-
peutical modalities of this
area of the skull base.
Experts from all over the
world describe their
experiences and give
an up-to-date report.

Springer-Verlag
Berlin
Heidelberg
New York
London
Paris
Tokyo
Hong Kong
Barcelona
Budapest